CONTROL TECHNOLOGY

pupils' follow-up book

Control Technology was written by G. J. Fox and D. F. Marshall, Danum Grammar School for Boys, Doncaster, and was edited for Project Technology by G. L. Viles. It was revised by P. W. Ghee, D. Hendley, A. Paul and G. L. Viles (editor).

Project
Technology

CONTROL TECHNOLOGY

pupils' follow-up book

HODDER AND STOUGHTON

LONDON   SYDNEY   AUCKLAND   TORONTO

British Library Cataloguing in Publication Data

Fox, G.J.
   Control technology.——2nd ed.
   Pupils' follow-up book
   1. Control theory
   I. Title II. Marshall, D.F. III. Viles, G.L.
   IV. Schools Council, *Project Technology*
   629.8   QA402.3

   ISBN 0 340 36407 6

First printed 1974
Eighth Impression 1985
Second edition 1986

Printed in Great Britain for Hodder and Stoughton Educational, a division of Hodder and Stoughton Ltd., Mill Road, Dunton Green, Sevenoaks, Kent by Richard Clay (The Chaucer Press) Ltd, Bungay, Suffolk

Contents – pupils' follow-up book

Foreword

Project Technology was a major curriculum-development project initiated by the Schools Council to promote a better understanding by boys and girls in school of the importance and relevance of technology. The Project was concerned with helping teachers to stimulate an *awareness* of the material and scientific forces with effect change in our society and to develop *knowledge* of these forces and their means of control by the direct involvement of pupils in technological activities.

The Project Technology teaching-material programme is the result of a careful assessment of what is required in the schools, followed by trials and editing of the material itself. The Project Technology team felt that it was essential to draw on the experience, imagination and flair of individual teachers who, over a period of years, had developed technological work in particular parts of the school curriculum.

It is against this background that the *Control Technology* course should be seen. Other teaching material, notably the Project Technology Handbooks series, indicates both a *thematic* and a *tactical* approach to school technology. All of this reflects the diverse nature of the work being done and of the alternative teaching methods and organisation which are possible.

This course, however, is intended to meet the needs of those schools who wish to develop a structured and sequential two- or three-year course, covering an important area of technology. Some teachers have expressed the view that, while appreciating the value of what might be regarded as the more fortuitous involvement of pupils with technological projects and investigations, they should welcome a more systemic course approach.

The joint authors of *Control Technology*, Messrs G. J. Fox and D. F. Marshall, have developed the course over a number of years at Danum Grammar School for Boys, Doncaster, where they were given essential support and encouragement by the Headmaster (Mr E. Semper, OBE) and by the Doncaster Education Authority.

Clearly defined educational objectives were established, and the appropriate teaching methods, based on pupil assignments, with appropriate texts and equipment, were progressively developed, with the support of the Project Technology team.

First-stage school trials were conducted in selected schools in the Doncaster area with the generous help of the LEA. Secondary trials followed in various parts of the country. We are especially grateful to the teachers in these schools for the help and cooperation they provided for the authors, the Project Technology team (led by Mr G. L. Viles) and the Evaluators appointed by the Schools Council.

The *Control Technology* course has been carefully based on purpose-built equipment essential to the effective running of the course. A considerable effort has been put into the development of this equipment, at all stages, and we are indebted to all concerned, including the present suppliers. More recently, Trent Polytechnic are to be thanked for encouraging further development work by the National Centre for School Technology.

This revised edition has been prepared to serve the current examination syllabus set by the Associated Examining Board and the Joint Matriculation Board who at the time of writing, are the examiners for this subject at 16 + .

Assignments using an electric motor have been amended for use with an improved model. Electronic assignments now follow current practise in circuit design, and the pneumatic component symbols now follow standards in current use.

Fluidic control is not included in current syllabuses and this original section has been removed. The logic section has been extended to meet current syllabus requirements and symbol notation.

Note
It is intended that these follow-up assignments will be used
individually at an appropriate time after the conclusion of the
relevant assignments.

Four or more continuously joined members cannot easily retain their shape under a load.

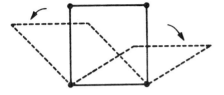 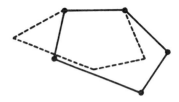

Such structures can be made rigid by using a number of methods.

a) Corner joints held by only one bolt or rivet (a pin-joint) can be made rigid by the addition of a gusset plate. This needs to be done at all joints to be most effective. This method is often used in buildings constructed of steel where openings must not be obstructed.

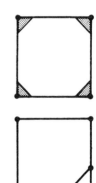

b) The corner joint(s) can be made rigid by bolting a short piece across the corner, thus forming a small triangle. This method is very similar to (a), and for good results must be applied to each joint. The longer the brace of a given shape and cross-sectional area, the more effective it becomes. (If you do not understand the meaning of cross-sectional area, ask your teacher.)

c) The simplest solution is to join either WY or XZ with an additional member. This now reduces the quadrilateral to two triangles. Triangulated structures are always rigid unless, of course, joints break or a side distorts.

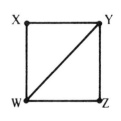

Which is the most effective solution — (a), (b), or (c)? Think about this and discuss the problem with other members of your group.

Any framework can be made rigid by reducing it to a number of triangles.

E.g.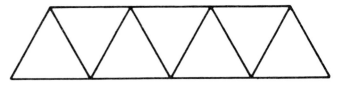

Consider the construction of electricity pylons, cranes, and bridge to see how this principle of triangulation is applied.

Structures Follow-up 2

The framework is unlikely to support the load satisfactorily.

a) The framework may fail because the vertical members may bend when the load is applied vertically.

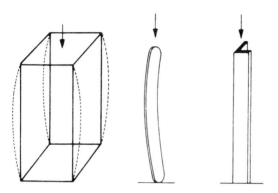

The vertical members are in compression.

When the members in a structure have to resist compressive forces, they must themselves be rigid to resist bending. A better section to use is a tube or 'I' section, but when using Meccano an angle girder will provide sufficient rigidity. Therefore you should have replaced the vertical strips with angle girders.

b) The framework may lean over and collapse if the force is not applied vertically *or* if the force is vertical but the structure is already leaning. The only force preventing leaning is the frictional force at the bolted joints.

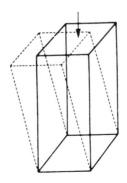

The solution to this problem should be clear to you — triangular bracing must be added to each side of the framework to provide the necessary rigidity.

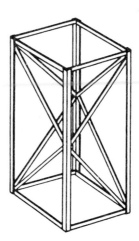

The final structure could be something like this. Angle girders should be used in each corner to resist the compressive forces, and the cross bracing must be employed to make the structure rigid, i.e. to resist forces which are not vertical.

The design of the tower can be further improved by increasing the size of the base as shown.

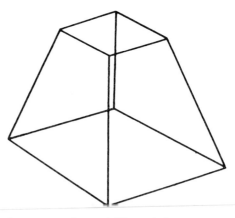

The larger base improves the stability of the structure (i.e. makes it less likely to topple over when tilted).

Structures Follow-up 3 _____

In designing this model road bridge you are faced with the same problem as the professional civil engineer, for in bridge building it is rarely possible to obtain structural members of sufficient length. Even if it were possible, their great weight or length would make them difficult and expensive to transport.

It is therefore necessary to design a framework which may be many times longer than the individual members, yet strong enough to support the specified loads.

To produce the necessary rigidity, the principle of triangulation is often used.

Some suitable frameworks are shown below.

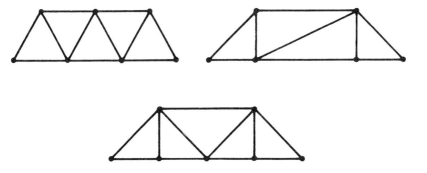

Of course, two similar frameworks would be required, suitably connected together to form a three-dimensional model.

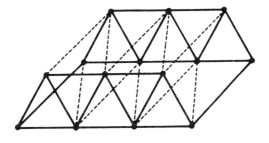

In a simple bridge, consisting of a single member supported at each end, the bridge as a whole will have to resist bending forces when loaded.

Due to the bending, the upper part of the member will be in compression and the lower part in tension.

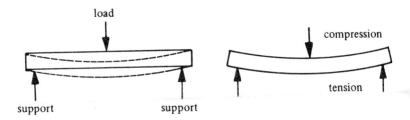

In practice, single members are rarely used; many members are joined together to form a structure of sufficient length.

To enable material to be used economically (a rigid angle girder is twice as heavy and therefore twice as expensive as a flat strip), it is necessary to determine the forces acting in a structure and to use angle girder only when essential (e.g. when a member is in compression).

NOTE: In pin-jointed structures members are either in tension or in compression if external forces are applied at the pin-jointed positions.

The diagram below indicates the forces present in a typical bridge structure caused by the loads applied at the pin joints.

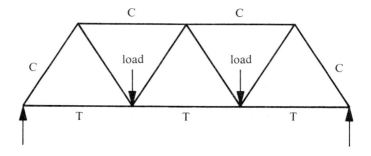

The sloping members of the framework would be in compression or tension as indicated. However, with the movement of a load, the forces on the four internal sloping members could change from tensile to compressive, therefore it is suggested that angle girders should be used for these members.

The example below shows a centrally placed load. The member actually supporting the load is in bending, but all other members are either in tension or in compression.

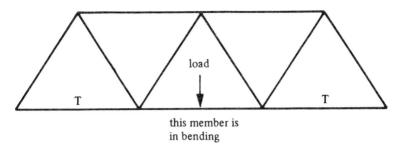

load

T   T

this member is
in bending

It should be noted that if a bridge, or part of a bridge, is supported anywhere other than at its ends, the framework will now bend in the opposite way, the upper part being in tension and the lower part in compression.

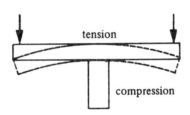

tension

compression

Any structure with an unsupported end is called a *cantilever*. Many motorway bridges are of this type, with a central, end-supported beam inserted between the cantilevers.

Can you suggest a reason for the shape of the beams?

You will have noticed on occasions how easily a measuring rule can be flexed, when a force is applied to its flat surface, compared with the almost impossible task of trying to bend it when applying a force on the 'edge' of the rule.

Beams have the property of being able to withstand greater bending forces when placed on edge, and this can be used to advantage. The beam on the left hand side of the diagram below has a cross sectional area of 12 mm². Stiffness depends upon the relative sizes of the breadth and depth of a beam. It is determined by the following formula

$$I = \frac{b \times d^3}{12}$$

where b is the breadth and d is the depth of the beam. The correct mathematical term for I is the second moment of area. However it is the magnitude (value) of this area that is of direct relevance to you.

Calculations for I are shown for the following beam cross sectional areas.

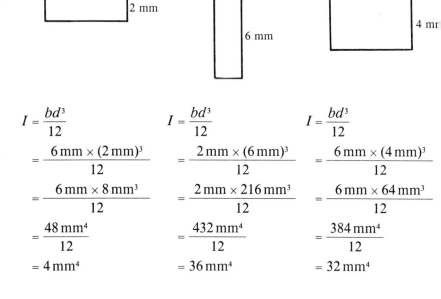

$$I = \frac{bd^3}{12}$$
$$= \frac{6\,mm \times (2\,mm)^3}{12}$$
$$= \frac{6\,mm \times 8\,mm^3}{12}$$
$$= \frac{48\,mm^4}{12}$$
$$= 4\,mm^4$$

$$I = \frac{bd^3}{12}$$
$$= \frac{2\,mm \times (6\,mm)^3}{12}$$
$$= \frac{2\,mm \times 216\,mm^3}{12}$$
$$= \frac{432\,mm^4}{12}$$
$$= 36\,mm^4$$

$$I = \frac{bd^3}{12}$$
$$= \frac{6\,mm \times (4\,mm)^3}{12}$$
$$= \frac{6\,mm \times 64\,mm^3}{12}$$
$$= \frac{384\,mm^4}{12}$$
$$= 32\,mm^4$$

It can be seen that the 6 mm × 2 mm beam when placed on edge can resist nine times the bending force than when it is placed down flat. It will also resist a greater bending force than a beam of twice the cross-sectional area (i.e. a 6 mm × 4 mm beam) placed down on its 6 mm side.

However when a beam is on edge, it can be deflected sideways by a small force, which may cause it to buckle and collapse under load. One solution is to use angle section, which is available in Meccano, to prevent buckling. A 'tee' section beam provides the same effect. Can you say why an angle section is practically more useful than a 'tee' section?

When a beam is resisting bending forces the compressive and tensile forces in the beam, referred to earlier, are greatest on the top and bottom surfaces. It is for this reason that the standard engineering Rolled Steel Joist has the bulk of its material placed at the top and bottom of a vertical I section as shown below.

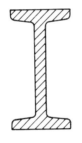

Structures Follow-up 4

As you will have discovered during the design stage, and by looking at the models made by other groups, this problem can be solved in a number of ways. This is true of most engineering problems — the final choice of design is determined by examining the cost, reliability, availability of materials, and other factors.

It is probable that most of the solutions fall into two standard forms:

a) the gantry type

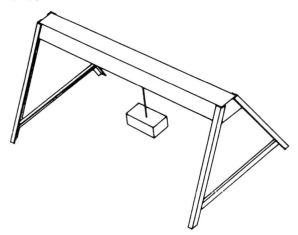

b) the cantilever type

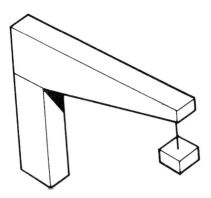

Provided that you used the principle learned in the previous lessons, you should have had no difficulty in producing a rigid structure

which would support the load without collapsing. You may have found that your structure was liable to twist when loaded, so attempt in your design to resist this *torsional* force.

Although your structure supported the load without distortion, your model may not have been satisfactory because it lacked stability, i.e. it easily toppled over.

This is more likely to happen in the case of the cantilever form of lifting device than in the gantry type, as explained by the following notes.

Stability

A free-standing structure is most stable if a vertical line drawn through its centre of gravity (if you do not know what this means, ask your teacher) falls well within the base area.

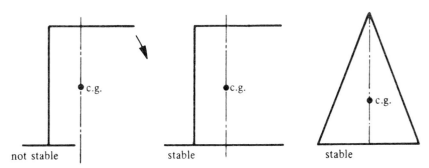

not stable   stable   stable

The gantry type

In the gantry type the c.g. can easily be kept within the base area and, provided the base is large enough, the structure will be stable.

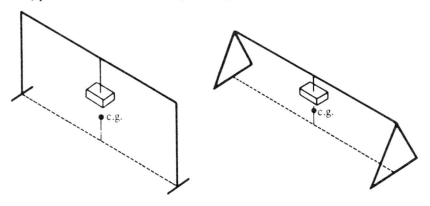

The cantilever type

To provide adequate stability, the base A must project beyond the
vertical line through the c.g.

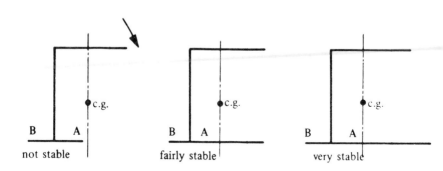

not stable   fairly stable   very stable

In the above diagrams, the part of the base marked B serves no
useful purpose, and can be left out of this kind of simple structure.

However, the base A may be an obstruction, in which case the
structure must be secured to the ground on the side B.

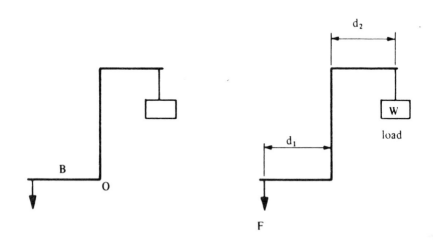

The securing point should ideally be as far from O as possible,
because when $d_1F = d_2W$ the structure becomes unstable.

Thus, the larger the distance d_1 becomes, the smaller F need be for a given distance d_2 and a given load W.

NOTE: When a load is being raised or lowered, the c.g. of the model shifts, and this must be allowed for in the design.

a) **Simple train of spur gears**

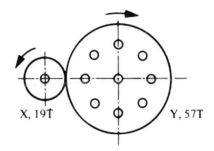

i) Gear X, 19 teeth; gear Y, 57 teeth — as indicated.
ii) Gears turn in opposite directions as shown.
iii) If gear Y is turned once, gear X rotates 3 times (57/19 = 3/1).
iv) The shaft of gear X is the easier to turn.

b) **Simple train of spur gears with an idler**

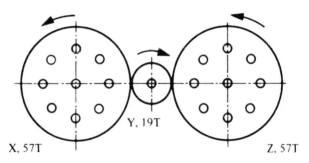

i) Gear X, 57 teeth; gear Y, 19 teeth; gear Z, 57 teeth — as indicated.

ii) The gears rotate as indicated, gear X in the same direction as gear Z.

iii) Gears X and Z rotate at the same speed. (Gear Y rotates 57/19 = 3 times if gear X is rotated once, but gear Z rotates 19/57 = 1/3 of a revolution for each revolution of Y.)

Gear Y is an idler. It has no effect on the gear ratio, but transmits the turning motion and reverses the direction of rotation of Z, compared with a system in which X drives Z directly.

iv) In this example, shaft X and shaft Z each offer the same resistance to turning.

Another example

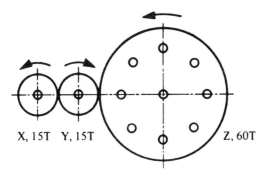

X, 15T   Y, 15T   Z, 60T

Gear X must rotate four times to rotate gear Z once. In this case, shaft X·will be easier to turn than shaft Z.

c) **Compound train of spur gears**

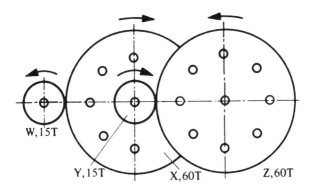

W, 15T

Y, 15T   X, 60T   Z, 60T

i)   Gear W, 15 teeth; gear X, 60 teeth; gear Z, 60 teeth.
ii) The gears rotate as indicated.
iii) If gear Z is rotated once, gear Y is rotated 60/15 = 4 times.
     Since gear Y is attached to gear X, gear X also rotates 4 times.
     If gear X rotates once, gear W rotates 60/15 = 4 times.
     Therefore, if gear X rotates 4 times, gear W rotates 4 × 4 = 16 times. Therefore the overall gear ratio of this system is 16:1, i.e. gear W must be rotated 16 times to rotate gear Z once.

By this means, using more gears if necessary, large gears ratios may be obtained.

iv) Gear shaft W is easily turned. Gear shaft XY is turned with difficulty. Gear shaft Z is even harder to turn. Why is this?

d) Spur gear and contrate gear

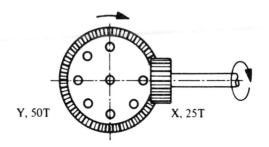

Y, 50T   X, 25T

i) Gear X, 25 teeth; gear Y, 50 teeth.

ii) The gears rotate in the directions indicated.

iii) This is basically the same system as in (a), the simple gear train, except that the gear shafts are mounted at right angles. Gear Y rotates once if gear X rotates twice (50/25 = 2/1).

iv) Gear X is the easier to turn. This gear system is not normally used in engineering, but it is a satisfactory method when using Meccano. The normal methods used by engineers to transmit motion through 90°, e.g. bevel and helical gears, are available in Meccano, but we shall not normally use them as their setting is more critical and the gears are more expensive.

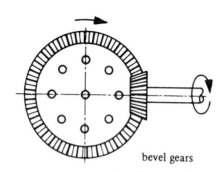

bevel gears

e) **Worm gear and pinion**

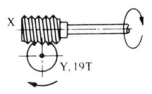

X

Y, 19T

i) Gear X is called the worm. It is rather like a screw-thread and acts as a one-tooth gear. Gear Y has 19 teeth. The shafts must be mounted at right angles to each other.

ii) The gears rotate as shown.

iii) As the worm gear X (effectively a one-tooth gear) is in mesh with a 19-tooth gear Y, it must rotate 19 times to turn gear Y once. This gear system is extremely effective when large-reduction gear ratios are required; e.g. a worm in mesh with a 200-tooth gear will give a reduction of 200:1.

iv) Gear X is easy to turn, but it is impossible to turn gear Y. This system is usually irreversible — the worm cannot be driven by the gear. This feature can be used to advantage in lifting devices. As the load is unable to turn the motor, a braking effect is obtained.

From these investigations the following generalisations can be made.

i) In spur-gear systems, adjacent gears rotate in opposite directions. (If the direction of rotation of one gear is reversed, the directions of rotation of all the other gears in the system are reversed.)

ii) The gear ratio of two adjacent gears is directly proportional to the number of teeth on the gears (a worm gear being equivalent to one tooth).

iii) Idler gears have no effect on the gear ratio, but they are used to:
   a) transmit movement from one gear to another,
   b) alter the direction of rotation of other gears.

iv) The gear which turns the most times (i.e. the smallest gear) is the easiest to turn.

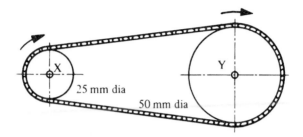

i)   Sprocket X is 25 mm diameter (18 teeth) and sprocket Y 50 mm diameter (36 teeth).
ii) Both sprockets turn in the same direction.
iii) Sprocket X turns twice to rotate sprocket Y once; 36/18 = 2/1.
iv) Shaft X is the easier to turn.

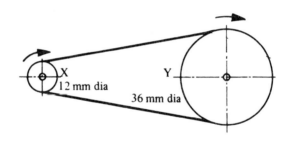

i)   Pulley X is 12 mm diameter and pulley Y 36 mm diameter.
ii) Both pulleys turn in the same direction.
iii) To rotate pulley Y once, pulley X must be turned approximately 3 times (a little more than 3 times).
iv) Shaft X is the easier to turn.

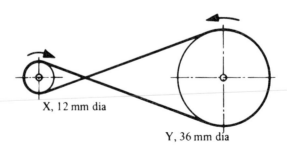

i) Pulley X is 12 mm diameter and pulley Y 36 mm diameter.
ii) The pulleys turn in opposite directions.
iii) To rotate pulley Y once, pulley X must be turned a little more than three times.
iv) Shaft X is the easier to turn.

The chain system transmits the rotary motion most positively, as the chain does not slip, because of the teeth on the sprockets.

In the pulley system, some slipping is bound to take place. The amount of slip will depend upon:
i) the load on the driven pulley;
ii) the tension in the belt;
iii) the arc of contact between belt and pulley. (The second system is better in this respect, as the belt is in contact with a greater part of the circumference of the pulleys.)

The main difference between these systems and those examined in *Gears 1* is that in gear systems the shafts have to be accurately positioned to give the correct meshing of the geasrs. In belt and chain systems the shafts can be any reasonable distance apart, because the chain and belt are adjustable in length.

The slipping in the belt-and-pulley system, although often a disadvantage, does build a certain degree of safety into some transmission systems. The belt will slip if overloaded, and this may prevent serious mechanical damage.

All gear, sprocket-and-chain, and pulley-and-belt systems are dangerous, and must be adequately guarded to prevent accidents.

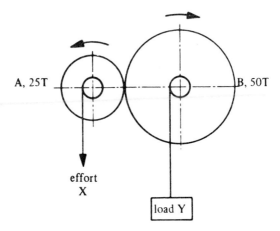

a) i) In this gear system, if the effort X moves 200 mm, the load
Y will be raised 100 mm. The reason for this is that the gear
A attached to the shaft supporting X will turn twice for each
revolution of the gear B attached to the axle supporting Y.

If A had 15 teeth and B had 60 teeth, X would move four
times as far as Y.

The ratio of the distance moved by X to the distance moved
by Y is the same as the gear ratio. Did you find this?

Note also that the turns ratio (the number of times A turns,
divided by the number of times B turns) is also the same as
the gear ratio:

turns ratio = gear ratio = distance ratio = 2
i.e. X moves twice the distance of Y.

ii) When you pulled X, you will probably have noticed that it
moved faster than Y.

Did you discover that the velocity of X was twice the velocity
of Y?

If you think about this carefully it must be so, since if X
moves twice as far as Y (and notice that they both start and
stop together, i.e. they move their respective distances in the
same time), X must move twice as fast as Y.

iii) In (i) we find that the distance ratio is 2 and in (ii) we find that the 'velocity ratio' is 2. It is easier to measure distances than velocities, so to find a velocity ratio we can say:

$$\text{velocity ratio (VR)} = \frac{\text{distance moved by effort}}{\text{distance moved by load}}$$

The distances moved by the parts of a machine are often important, so a knowledge of its velocity ratio can be useful.

b) Just as distance ratios are important, so are 'force ratios'. In this experiment, in which you lifted 1 kg and noted the force required as indicated on a spring-balance, you found the force ratio (sometimes called the mechanical advantage) when you divided the load by the effort:

i.e. $$\text{force ratio (mechanical advantage, MA)} = \frac{\text{load}}{\text{effort}}$$

c) i)   You will have found that the mechanical advantage of the machine using nylon bearings is greater than that using plain bearings. Nylon bearings give a higher efficiency than the plain bearings because there is less friction. Did you suggest friction in 3(ii)?

$$\text{Efficiency} = \frac{\text{useful work got out of machine}}{\text{work put into machine}} \times 100\%$$

Efficiency has a definite meaning, and is usually expressed as above.

You yourself put the work into the machine by pulling on the spring-balance, and your effort is measured by the force indicated on the spring-balance.

The work put into the machine is given by:

effort $\times$ distance moved by effort

If the effort is F newtons, and the distance it moves is x metres, then:

work put into machine $= F \times x$ newton metres
$= Fx$ joules (J)

since 1 newton metre equals 1 joule.

The useful work got out of the machine is given by:
load lifted $\times$ distance load is lifted

If the load is W newtons, and it is lifted through y metres, then:

work got out of machine $= W \times y$ newton metres

$= Wy$ joules

i.e. Efficiency $= \dfrac{\text{load} \times \text{load distance}}{\text{effort} \times \text{effort distance}} \times 100\%$

In this case, the load is the gravitational force acting on 1 kg, i.e. approximately 10 newtons.

ii) Again, just as velocities are difficult to measure, it is often inconvenient to measure distance moved by the parts of the machine. If you know the VR and the MA of a machine, you can find the machine's efficiency, even without seeing it, as shown below:

Efficiency $= \dfrac{\text{useful work got out of machine}}{\text{work put into machine}} \times 100\%$

$= \dfrac{\text{load} \times \text{load distance}}{\text{effort} \times \text{effort distance}} \times 100\%$

$= \dfrac{\text{load}}{\text{effort}} \times \dfrac{\text{load distance}}{\text{effort distance}} \times 100\%$

But $\dfrac{\text{load}}{\text{effort}} = \text{MA}$

and $\dfrac{\text{load distance}}{\text{effort distance}} = \dfrac{1}{\text{VR}}$ (since $\text{VR} = \dfrac{\text{effort distance}}{\text{load distance}}$)

therefore efficiency $= \dfrac{\text{MA}}{\text{VR}} \times 100\%$

iii) *Calculate the efficiency of the two gear systems in your notebook. How near to 100% is the efficiency?*

iv) Assuming that the efficiency of the machine were 100% (this is impossible in practice, since there would be no friction), what could you then say about the mechanical advantage and velocity ratio?

When you have arrived at an answer, *complete the following statement in your notebook:*

'For a machine with an efficiency of 100%, the following is true ——.'

Select (a) the MA is greater than the VR,
or   (b) the VR is the same as the MA,
or   (c) the VR is greater than the MA,
or   (d) the VR is less than the MA.

_____

Because both the voltage and the current were kept constant, the power input to the motor was the same for all the experiments, and therefore a valid comparison of the lifting capabilities of each gear ratio can be made.

(Power = rate of working, or the work done per second. Avoid using the word 'power' unless you mean the rate of working.)

When plotted on your graph, the results of gear ratio and load lifted should approximate to a straight line passing through the origin of the graph. This indicates that the two variables are directly proportional to each other. If the gear ratio is doubled (i.e. the speed of rotation of the shaft is halved), the load lifted is doubled.

It will have been noticed, that when the gear ratio is doubled, the load lifted is doubled, but the velocity at which the load is lifted is halved. (In the same time it moves through half the distance.)

If W newtons is the load, and x metres the distance it is lifted when using the 18:1 gear ratio, the work done on the load is given by:

$$\text{load} \times \text{distance load moves} = W \times x \text{ newton metres}$$
$$= Wx \text{ joules (J)}$$

since 1 newton metre equals 1 joule.

The rate at which work is done is given by

$$\frac{\text{work done (in joules)}}{\text{time taken (in seconds)}} = \text{rate at which work is done (in watts)}$$

A practical example

Ratio 6:1
Load lifted, 2 newtons; distance lifted, 0.5 m; time, 3.5 seconds.
Work done = 2 × 0.5 = 1 joule
Rate at which work is done $= \dfrac{1}{3.5} = 0.285$ watts

Ratio 12:1
Load lifted, 4 newtons; distance lifted, 0.5 m; time, 7 seconds.
Work done = 4 × 0.5 = 2 joules
Rate at which work is done $= \dfrac{2}{7} = 0.285$ watts

This shows that the work done is different but that the rate at which work is done is the same. Therefore, no matter what gear ratio you use, in a given time the same work is done because an increase in load always results in a proportional decrease in velocity.

You were warned that it was essential to prevent the string from winding over a previous layer as this would increase the effective diameter of the shaft and the load would be lifted more quickly.

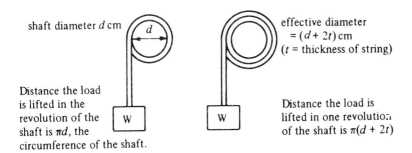

shaft diameter d cm

effective diameter
= $(d + 2t)$ cm
(t = thickness of string)

Distance the load is lifted in the revolution of the shaft is πd, the circumference of the shaft.

Distance the load is lifted in one revolution of the shaft is $\pi(d + 2t)$

If this happens, the work done in lifting the load will be increased (work done = load × the distance it is lifted), more power must be provided by the motor, and the current in the circuit will therefore rise.

Did you notice that, when the load was lowered, the current used by the motor was much less than that used when the weight was lifted? Think about this.

You may have noticed that the voltage from the power-supply unit rises when the motor is switched off. At the moment it is unlikely that you will be able to explain this. Later you will discover the reason for the effect, but, until then, keep in mind that the voltage across the terminals of a power supply changes if the current flowing changes.

Basic Electricity Follow-up 1

a) When you reversed the cell you should have found that the bulb still lit normally. For the moment we can liken electricity to a kind of liquid which flows from the positive terminal of a battery to the negative one. When you reversed the cell, therefore, you simply caused the electricity to flow in the opposite direction round the circuit.

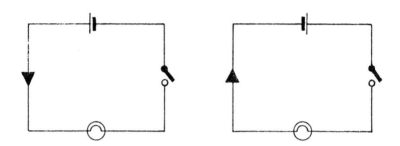

b) With two cells you will have found that the current taken by the bulb is doubled, or nearly so. With three cells the current should treble, or nearly so.

The brightness of each bulb increases if two cells are used, and with three cells it may become so bright that the bulb filament burns out. The higher the current in a bulb filament, the hotter it becomes, and the greater the light emitted by it.

c) **Parallel arrangement**

The bulbs should be found to have about normal brightness.

You can find the current consumed by each bulb by connecting the ammeter as shown:

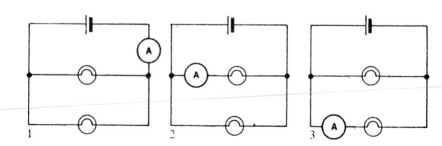

The current taken by each bulb is about the same. The ammeter in diagram 1 should read about twice as much as the ammeters in diagrams 2 and 3. The current in a parallel circuit divides into two parts at A and joins up again at B to return to the cell as shown in diagram 4.

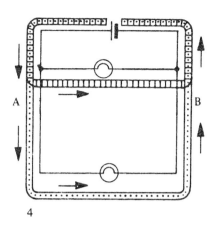

4

d) **Series arrangement**

The bulbs will be dimly lit. The current taken by each bulb is the same, and can be measured by placing the ammeter at any point in the circuit. The current is less than that taken by a bulb of normal brightness. Why do you think this is so?

e) Your circuit should look something like this:

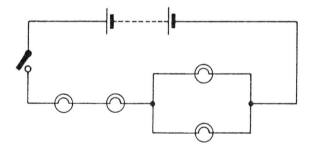

Basic Electricity Follow-up 2

a) When a cell is used in a circuit, it 'forces' electricity around the circuit, rather like water is forced through a water-pipe by a pump. An ammeter does *not* measure the total amount of electricity which flows in a circuit, otherwise the reading on your ammeter would continuously increase. The ammeter measures the *rate* at which electricity flows. This rate is called a current, and we measure it in amperes. Measuring the rate of flow of electricity is not unlike measuring water flowing in a pipe, in litres per minute or litres per second. Similarly, we could count the *total number* of spectators passing through the turnstile at a football match, or we could count the *rate* at which they pass through. A meter which measured the rate at which spectators pass would be acting like an ammeter.

When measuring a current in a circuit, we simply break the wire at some point and insert the ammeter. Should you attempt to fit an ammeter in any other way, you will be connecting it incorrectly, and you may damage the ammeter or some other component in the circuit.

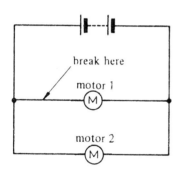 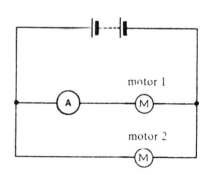

b) **Cells in series**

When two identical cells are connected in series, the electrical force being applied to the circuit is doubled; e.g. if you connect two 1.5 volt cells in series, you get a force of 1.5 + 1.5 = 3.0 volts.

If the cells are different, you simply add up the individual cell voltages; e.g. if a 2 volt cell is connected in series with a 1.5 volt cell, you get a electrical force of 2 + 1.5 = 3.5 volts.

c) Your results should agree with (b) above. In effect your voltmeter is a counter of cells in series. In future, you can use any *direct-current* source of electricity, and to find out what force it represents you can simply connect a voltmeter across the supply terminals. (A cell always provides direct current.)

d) **Cells in parallel**

When two identical cells are connected in parallel, the force is still the same as a single cell, but the pair can provide a given current for twice as long.

e) The cells in a car battery have an electrical force of about 2 volts each. They are connected in series to form a battery of 6 cells having a total voltage of $6 \times 2 = 12$ volts.

Basic Electricity Follow-up 3

a) i) Resistance wires are made from special *alloys* for applications such as electric-fire elements, immersion-heater elements, etc. Typical alloys are nichrome, eureka, contra and manganin. You may wish to find out what metals these wires contain.

   ii) You should have found out that, when the resistance in a circuit is *doubled*, the current is *halved* if the applied voltage is the same throughout. Another way of expressing this is:

   the current is inversely proportion to the resistance

   or $I \propto \dfrac{1}{R}$ (I is the current and R the resistance)

b) In this experiment, if the wire does not become unduly hot, you will no doubt have found that, if the voltage is *doubled*, then the current is *doubled*. Another way of expressing this is:

   the current is directly proportional to the applied voltage

   or $I \propto V$ (V is the voltage)

c) Have you found a rule which always applies?

   We can combine $I \propto \dfrac{1}{R}$ and $I \propto V$ to give: $I \propto \dfrac{V}{R}$

   and if we choose suitable units for voltage, current, and resistance, we can say:

   $I = \dfrac{V}{R}$ (This is very important.)

   Suitable units to use are:

   current . . . . amperes
   voltage . . . . volts
   resistance . . ohms

   You may not have met the unit *ohm* (symbol Ω). It is the resistance of a material if a voltage across it of one volt produces a current through it of one ampere.

   Example What is the resistance in a circuit if the applied voltage is 10 volts and a current of 2 A flows?

   $I = \dfrac{V}{R}$

   $\therefore V = IR$ or $R = \dfrac{V}{I}$ $\therefore R = \dfrac{10}{2} = 5$ ohms

Electrical Switching Follow-up 1

There are a number of ways of finding the speed of your motor output shaft. Here are some examples:

a) Put your motor on its greatest gear reduction (360:1). Make a chalk mark at some point on the output shaft. Count the number of revolutions the motor output shaft makes, over a fairly long period of time.

Example
Gear ratio = 360:1
Number of revolutions counted = 50
Time for the number of revolutions = 3 min

$$\text{Speed of output shaft at 360:1} = \frac{50}{3} \text{ rev/min}$$

$$= 16.66 \text{ rev/min}$$

Since the gear ratio is 360:1, the basic motor speed is 360 × 16.66 = 6000 rev/min.

To find a motor output shaft speed using any gear ratio
Assuming you wish to use the motor on a 12:1 reduction, the output shaft speed will be:

$$\frac{\text{basic motor speed}}{\text{gear ratio}} = \frac{6000}{12}$$

$$= 500 \text{ rev/min}$$

Selecting the correct gear ratio
Alternatively, if you want an output shaft speed of 333 rev/min, you will need a gear ratio of:

$$\frac{\text{basic motor speed}}{\text{output shaft speed}} = \frac{6000}{333}$$

$$= 18.02$$

nearest available ratio = 18:1

b) Cut out a disc of card about 12 cm in diameter, and fit it centrally onto the motor shaft. Make a clear mark near the circumference. Count the number of times the mark goes past a given point in one minute.

You will find that this method enables you to count the revolutions of the shaft more easily than making a mark directly on the shaft.

NOTE: Both the above methods are suitable for slow shaft speeds. For rather faster moving shafts, the following method can be used. (Ask your teacher for more details, beyond those shown below, and for the necessary apparatus.)

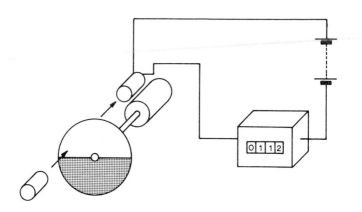

A disc of clear plastic is needed, about 15 cm in diameter, with half of it covered with tape or painted. The disc is fitted onto the motor shaft. A photocell is mounted on one side of the disc and a light source on the other. The photocell is connected to a battery and an electromagnetic counter. Each time the clear half appears between the cell and the light source, a count of one appears on the counter. With a suitable electromagnetic counter, rates of up to 25 per second are possible, which represents a motor speed of 1500 rev/min.

Photocells are also referred to as 'light-dependent resistors'.

The above circuit may not work satisfactorily at high speed. Your teacher may suggest a simple amplifying circuit.

a) i) With the switch 'UP' the motor runs, using pairs 2 and 3 or 5 and 6.

This shows that in the 'UP' position of the switch, sockets **2 and 3** and **5 and 6** are **shorted out**. Sockets **1 and 2** and **4 and 5** remain **open-circuited**.

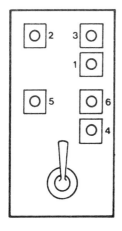 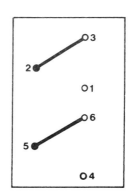

ii) With the switch 'DOWN' the motor runs, using pairs **1 and 2** and **4 and 5**.

This shows that in the 'DOWN' position of the switch, sockets **1 and 2** and **4 and 5** are **short-circuited**. Sockets **2 and 3** and **5 and 6** remain **open-circuited**.

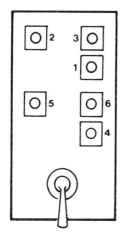 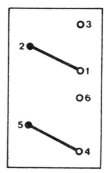

If you look carefully at the diagrams above, you will see that the unit consists of two switches. One switch has contacts 1, 2, and 3; and the other has contacts 4, 5, and 6. The switches are change-over switches, contact 2 connecting to either 1 or 3, and contact 5 joining to 4 or 6.

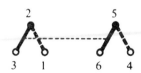

The broken lines show the alternative switch position. Contacts 2 and 5 are used in both position, and are called the 'poles' of the switches. (Note that the circle indicating the pole is shaded in.)

The dashed line indicates that the two switches are 'ganged', i.e. they both change over together.

Your switch unit contains a *two-pole change-over* switch (sometimes called two-pole, two-way).

Other kinds of switches

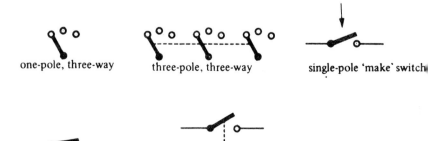

one-pole, three-way

three-pole, three-way

single-pole 'make' switch

single-pole 'break' switch

two-pole 'make' switch
two-pole on–off

b) The circuit below shows a method which completely isolates the motor from the supply.

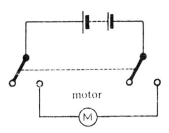

motor

Your switch is used as a two-pole on-off.

c) If the supply leads are interchanged, the motor shaft rotates in the other direction.

d) A circuit for reversing the motor can be connected in this way:

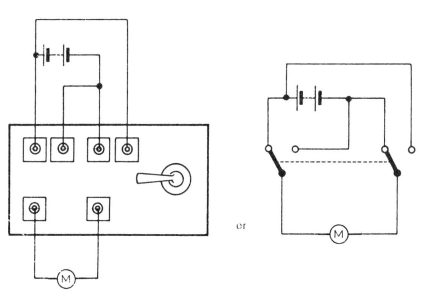

or

e) There are at least two possible ways of using the switchboxes to make the motor run forwards, in reverse, and to stop it.
   i) This circuit uses a two-pole change-over switch and a single-pole on-off switch:

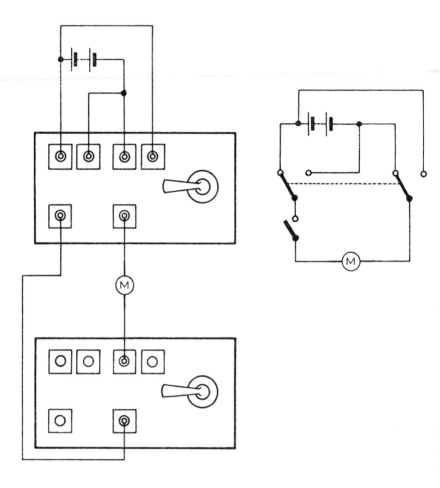

ii) This circuit uses two *separate*, single-pole change-over
switches.

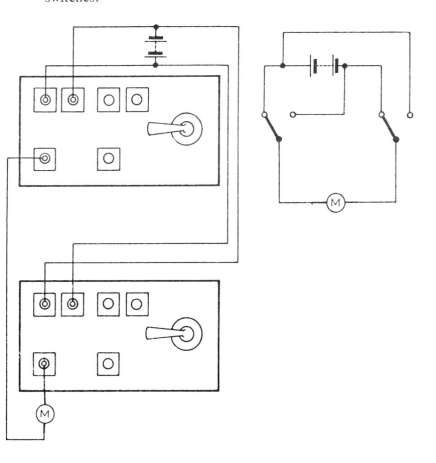

IMPORTANT: Make sure that you have tried both these circuits.
If you replace your motor with a bulb, you will see that this
switching arrangement is the same as that used on a stairway.

a) Your microswitch is a single-pole change-over type.
   Microswitches are useful in controlling machines and
   electromechanical devices, since they enable switching to be
   carried out using only small forces.

   Your teacher will show you other kinds of microswitches if you
   ask to see them.

   The switch may be normally-open, normally-closed, one-pole
   change-over, or two-pole change-over.

   Your circuit should look like this:

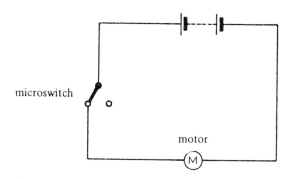

b) The reed-switch is a normally-open type. Your circuit should
   look like this:

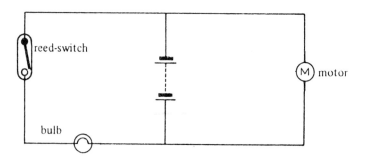

c) If you set up the circuit below, and place a magnet on your vehicle near to the reed-switch, the motor will run.

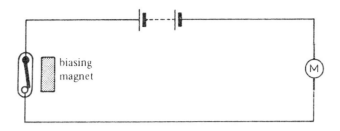

When the reed switch approaches the magnet on the obstruction, the reed-switch opens, provided that the two magnets have opposite poles facing each other. The reed-switch lies in a region of low magnetic field strength, and hence it opens.

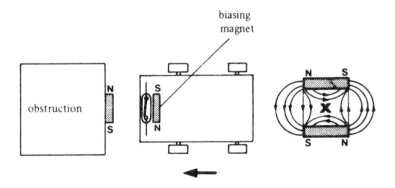

'X' is a neutral point in the magnetic field where the resultant magnetic field strength is zero.

Electrical Switching Follow-up 4

a) When no plugs are fitted into the green sockets (no current flows in the relay coil), sockets 2 and 3 are connected and 5 and 6 are connected.

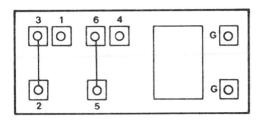

When the green sockets are connected to the 12 V supply the relay *'energises'*, and sockets 1 and 2 and 5 and 4 are connected together:

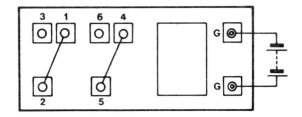

You should now have realised that the relay switch is a two-pole change-over type, the same as in your switchbox. In the relay-box, an electro-magnet is used to change over the connections, instead of a lever (or 'toggle' as it is sometimes called — lever switches are often called 'toggle switches').

The symbol for a relay is:

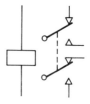

The type and number of contacts depend upon the actual relay.

b) Your arrangement should look something like this:

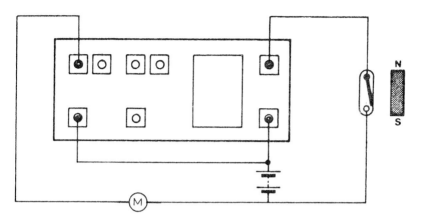

Now draw this circuit diagram in your notebook, unless you have already done so.

c) The arrangement below will enable your vehicle to reverse away from the magnet fitted to the obstruction:

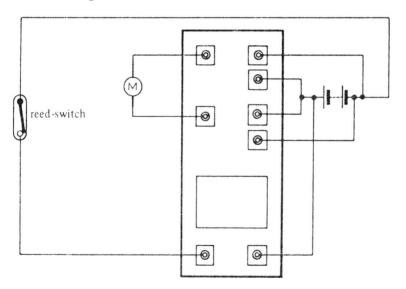

41

You will notice that the vehicle reverses away from the magnet on the obstruction, but only for a short distance. It then changes direction and moves forward toward the magnet again. This will continue indefinitely. We say that the vehicle '*oscillates*' backwards and forwards. What other kinds of oscillation (to-and-fro motion) can you think of?

This oscillation can be regarded as being between two 'states' (forwards and backwards). However, if the magnet is removed the vehicle will no longer oscillate but will move forwards only; i.e. forward is the preferred state.

The preferred (forwards) state can be regarded as stable and the reverse state as unstable. Such an arrangement or device is known as a *monostable* device, i.e. having two states only one of which is stable.

However, if we consider a vehicle that always has a magnet to influence its movement, it would oscillate continuously. Any such arrangement or device is known as an *astable* device; i.e. having two states neither of which is stable.

You will use a third relay system, or switching device, later in the course which can remain stable in either state indefinately, until it is switched to change state by an external signal. Such a device is known as a *bistable* device; i.e. having two stable states.

Now draw a circuit diagram of your reversing arrangement in your notebook, if you have not already done so.

a) Your circuit should look like this:

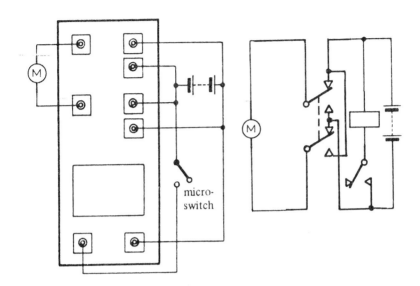

micro-
switch

Your vehicle will *oscillate*, but very quickly, so that it barely
leaves the obstruction before it again reverses back into it.

b) When the relay is connected to the supply it *immediately* closes
(energises). On removing one of the supply leads, the relay
immediately opens (de-energises).

c) With a capacitor connected between the relay-unit green sockets
(connected across the relay coil), the relay energises immediately
it is connected to a supply. On disconnecting the supply,
however, there is a *delay* of 3 or 4 seconds before the relay de-
energises.

The capacitor behaves as a reservoir for electrical charge. On
connecting it to a supply it charges, i.e. it stores some charge.
When the supply is disconnected, the capacitor 'discharges'
through the relay coil, i.e. current flows through the coil for
some time, even though the supply has been disconnected.

Eventually you will be studying the capacitor more fully. For the
present, however, you should be aware that a capacitor and
relay can be used to produce a *delay*.

d) By fitting a capacitor across the relay coil, you have introduced a delay. Your vehicle should, therefore, reverse away from the obstruction for half a metre or so before it automatically reverses itself again and once more moves towards the obstruction.

Your circuit should look like the one shown in (a) above, but with a capacitor included across the relay coil.

e) i) You should have found that two capacitors of equal value connected in parallel have an effect roughly equal to that of a capacitor of twice their individual value. In parallel, add up the individual capacitor values, e.g. 2000 μF in parallel in 4000 μF is equivalent to 6000 μF.

ii) You should also have found that two capacitors of equal value connected in series store half as much charge as one of them used on its own; i.e. two 5000 μF capacitors in series are equivalent to a 2500 μF capacitor.

Electrical Switching Follow-up 6 _____

a) The double-relay unit is called a 'flip-flop' or 'bistable'. Do you think these names are appropriate?

b) When the bistable is first connected to the supply, nothing appears to happen.

c) When a pair of green sockets is short-circuited, you will have noticed that one relay energises. It remains energised even if the short-circuit is removed, i.e. an *'instantaneous'* short is sufficient to cause the relay to close and remain closed.

Because the unit can 'memorise' the fact that a short-circuit has been produced across the green sockets, even after the short has been removed, it is classed as a *memory device*.

d) A short across the second pair of green sockets causes one relay to energise and the other to de-energise, i.e. the unit 'flips' over from one state to another.

e) Your table should look like this:

	sockets joined together
Relay A energised	5 and 4
Relay B de-energised	2 and 3
Relay A de-energised	2 and 1
Relay B energised	5 and 6

f) Your circuit layout should look something like this:

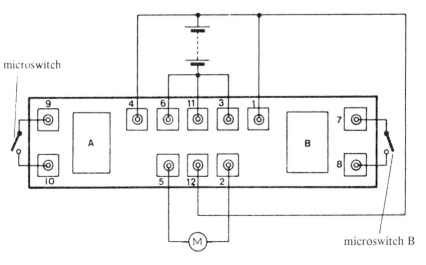

The circuit diagram for the above layout is:

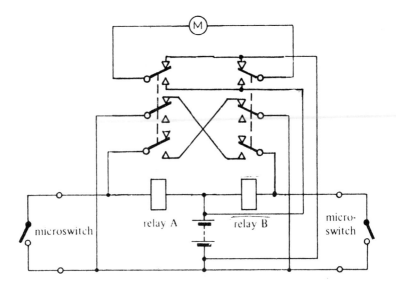

After microswitch A has been operated (and released) the circuit will become

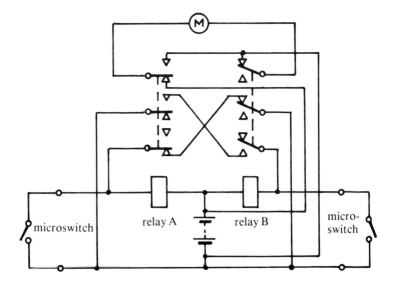

A somewhat simpler but less versatile bistable device can be constructed with a single relay unit.

Study the circuit diagram:

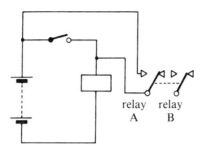

When the microswitch is operated the relay energises, contacts RLa close and so short-circuit the microswitch. Thus even when the microswitch is released the relay remains 'latched' on.

This circuit is of limited use as the only way to unlatch the relay is to disconnect the power supply. However if a second normally-closed switch is included as shown below then the circuit can be reset by operating this switch.

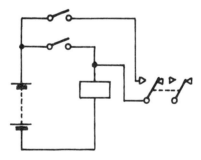

Of course this circuit only gives you access to the remaining single-pole change-over contacts (RLb) on the relay.

Your teacher may have some four-pole change-over relay units which you can use if the need arises.

a) When light from lamp 1 strikes the photocell, lamp 2 glows if the illumination is sufficiently intense. As the distance between the light source (lamp 1) and the photocell is increased, lamp 2 becomes dimmer.

The photocell you have used is a light-sensitive resistor (often called a 'light-dependent resistor', or 'l.d.r.'). When fully illuminated, its resistance is very low, but in darkness the cell resistance is very high (several hundred thousand ohms). The actual values vary to some extent from cell to cell. In your experiment, you *decreased* the illumination of the cell by *increasing* the distance between it and the light source.

b) i) When light is incident upon the photocell the relay is energised.

   ii) Interrupting the beam of light causes the relay to de-energise.

   iii) The relay should operate reliably up to a distance of at least 450 mm. At greater distances the combined effect of the smaller amount of light and the very critical alignment between the photocell and the light source results in less reliable operation.

If the amount of light hitting the cell is small, the resistance of the photocell limits the current in the circuit to such a value that it is insufficient to energise the relay fully.

Read again the 'Warning' given in the assignment.

Electrical Switching Follow-up 8

a) Oscillating circuit using a photocell

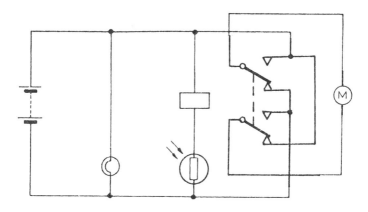

You should have had no difficulty with this circuit, shown above, as it is a combination of those used in previous assignments.

b) The behaviour of your vehicle with a capacitor connected across the relay coil depends on a number of factors:

the design of your vehicle,

the position of the photocell and the light source,

the size of the capacitor.

Typical behaviour if the light beam to the photocell is interrupted by the whole length of the structure

Small capacitor:

i) Photocell illuminated. Relay energised. Capacitor charged.

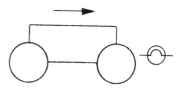

ii) Light beam to the photocell interrupted. Relay held on by the capacitor.

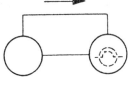

iii) Capacitor gradually discharges, relay changes over (de-energises) after short delay. Vehicle reverses.

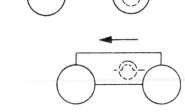

iv) Photocell illuminated. Relay energised. Vehicle moves forward again.

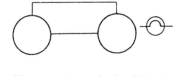

In these circumstances, the vehicle will pass through the light beam and continue some distance before reversing and moving forward again.

In position (i), with the vehicle moving forward, the photocell is illuminated, the relay coil is energised, and the capacitor is charged.

In position (ii) the photocell (now not illuminated) cuts off the power supply to the relay coil, but the capacitor discharges through the relay coil to keep it energised for a further period, so that the vehicle continues forward.

In position (iii) the capacitor, now almost discharged, can no longer 'hold on' the relay, so the relay contacts change over and the vehicle reverses.

When the photocell is again illuminated (position (iv)), the relay is energised, the capacitor is charged, and the sequence is repeated.

Large capacitor:

i) Photocell illuminated. Relay energised. Capacitor charged.

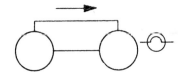

ii) Light beam to the photocell interrupted. Relay held on by the capacitor.

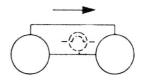

iii) Relay still held on by the capacitor. Photocell illuminated. Relay energised by the battery through the photocell. Vehicle continues forward.

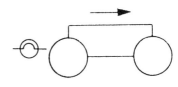

In this case, the capacitor holds the relay on while the vehicle passes through the light beam, and, as soon as the photocell is illuminated again, the relay is energised from the battery and the vehicle continues forward.

Typical behaviour if the light beam is interrupted for a short period only (e.g. by the wheels or only part of the structure)

i) Photocell illuminated. Relay energised. Capacitor charged.

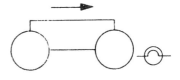

ii) Light beam to photocell interrupted. Relay held on by capacitor.

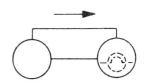

iii) Photocell again illuminated. Relay energised.

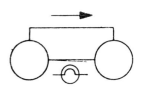

In these circumstances, the vehicle will probably continue to move forward, passing completely through the light beam.

In position (i), with the vehicle moving forward, the photocell is illuminated, the relay coil is energised, and the capacitor is charged.

In position (ii) the photocell (now not illuminated) cuts off the power supply to the relay coil, but the capacitor discharges through the relay coil to keep it energised for a further period, and the vehicle continues forward.

If the capacitor produces a long enough delay to drive the vehicle to position (iii), the relay coil again becomes energised from the supply and the vehicle continues forward.

A similar action will take place when the second pair of wheels passes.

If the capacitor discharges before position (iii) is reached, the relay will change over and the vehicle will reverse, but will move forward again as soon as the photocell is illuminated as shown below.

i) Photocell illuminated. Relay energised. Capacitor charged.

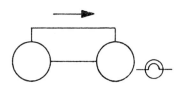

ii) Light beam to photocell interrupted. Relay held on by the capacitor.

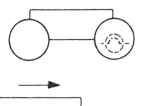

iii) Capacitor discharges. Relay changes over. Vehicle reverses.

iv) Photocell illuminated. Relay energised. Vehicle moves forward again.

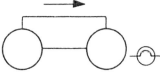

c) You probably expected your vehicle to behave as it did in the previous 'delay' circuit using a microswitch and capacitor, i.e. reverse for a short distance on breaking the beam of light and then move forward again.

The reason for the difference in behaviour is that, when using a photocell to control the relay, the vehicle moves *forward* when the relay coil is *energised* and *backwards* when *de-energised*. In the 'delay' circuit previously used, the forward motion was produced when the relay coil was de-energised.

NOTE: When using a photocell to drive a relay, the relay coil will be energised if the photocell is illuminated.

If this condition is not acceptable, a second relay can be used to 'invert' the output from the photocell. (Do you remember using a relay to 'close' the normally-open contacts of a reed-switch?)

A circuit using a light source, a photocell, two relay-units, and a capacitor to make a vehicle reverse for a distance when it interrupts a beam of light

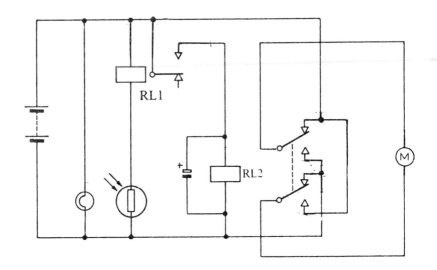

b) When the light beam is interrupted, RL1 is de-energised, causing RL2 to be energised and the capacitor to be quickly charged directly from the supply. The vehicle moves backwards.

RL1 de-energised. RL2 energised.
Capacitor charged. Vehicle moves
backwards.

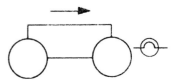

a) When the photocell in the above circuit is illuminated, relay 1 is energised and relay 2 is de-energised. In this condition, the vehicle moves forward towards the light beam.

RL1 energised. RL2 de-energised.
Vehicle moves forward.

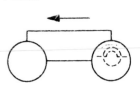

c) The vehicle reverses out of the light beam. RL1 again becomes energised, removing the power supply to RL2, but it remains energised as the capacitor *slowly* discharges through the relay coil. The vehicle continues to move backwards.

RL1 energised. RL2 disconnected from the supply but energised by the capacitor. Vehicle continues backwards.

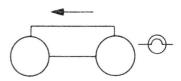

d) When the capacitor is sufficiently discharged, RL2 changes over, and the vehicle moves forward until it again interrupts the light beam to the photocell.

Capacitor discharged. RL2 de-energised. Vehicle moves forward again.

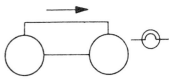

Electrical Switching Follow-up 10

a) i) The bistable relay unit does not function satisfactorily if both of the photocells are normally illuminated.

   When illuminated photocells are connected across each pair of green sockets, both relays will be energised. If one photocell is blanked off, the relay adjacent to its pair of green sockets will de-energise, but will become energised again as soon as illumination is restored. If both photocells are blanked off simultaneously, both relays will become de-energised, but in practice one relay will probably remain energised (its coil being fed through the contacts of the other de-energised relay) because it is difficult to blank off both cells at exactly the same time.

   Bistable devices are normally designed to accept pulse inputs and, using the bistable relay unit, the pulse is a momentary short-circuit across either pair of green sockets.

   ii) When both of the photocells are normally blanked off, the bistable relay unit will function normally provided only one cell at a time is illuminated.

b) To make the bistable relay unit function satisfactorily when the photocells are normally illuminated, a further relay is required for each photocell, to 'invert' its output.

Open relay contacts (they close only when the photocell is blanked off):

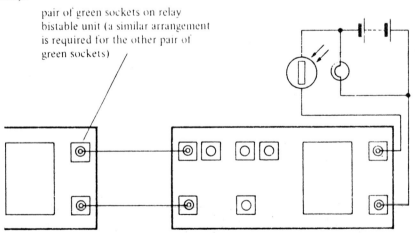

pair of green sockets on relay bistable unit (a similar arrangement is required for the other pair of green sockets)

c) Your circuit layout should look something like this:

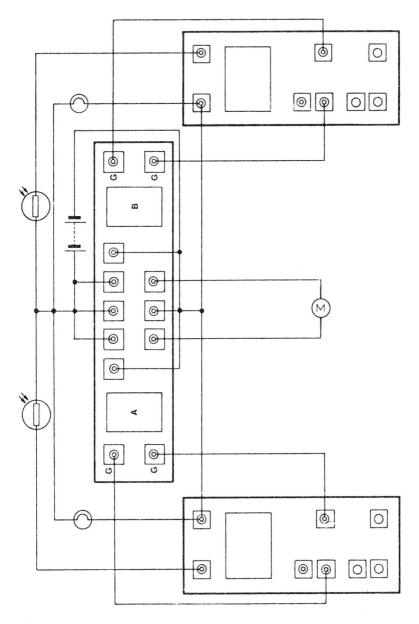

It should be noted that the motor will not run until *one* photocell is obscured.

The complete circuit diagram, using two photocells, two relay-units and a relay bistable unit to control your vehicle

KEY: G = green socket   B = blue socket
     R = red socket

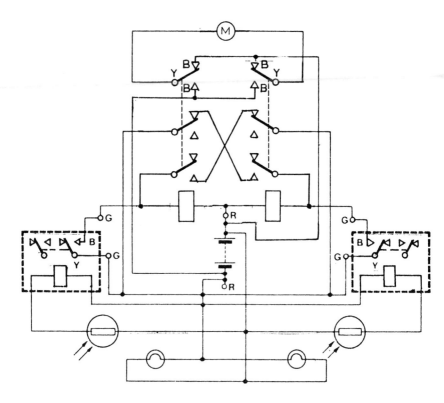

It should be noted that the motor will not run until *one* photocell is obscured.

Linear Motion Follow-up 1

There are many devices for converting rotary motion to linear motion. Your teacher will have shown you some important ones. Some commonly used methods are:

a) **The crank and slider**

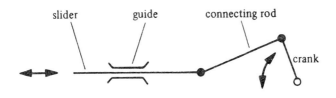

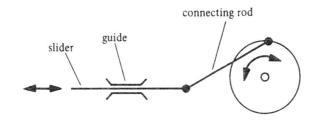

Properties
i)   Large push and pull forces can be exerted.
ii) The distance the slider moves is determined by the length of the crank.
iii) This mechanism is useful for large linear movements.
iv) Since the crank moves in a circle, only one kind of motion of the slider is possible. (Is the motion fast-slow-fast, or what? Try to work this out starting with the crank in line with the slider for one revolution.)
v)   The motion of the slider is reciprocating (backwards and forwards).

Applications
The piston connecting rod and crankshaft in a motor car.

b) The peg and slot

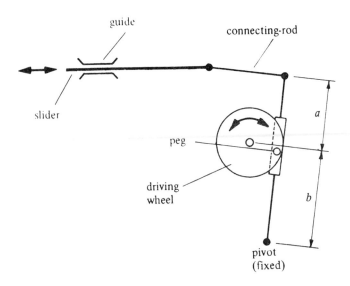

Properties

i) The distance the slider moves depends upon:
   a) the distance of the peg from the centre of the driving wheel, and
   b) the ratio of distance *a* to *b*.

ii) This mechanism is used to produce large to very large linear movement.

iii) Large push and pull forces can be exerted, but one is always greater than the other. In the diagram shown below, the push force is greater than the pull force.

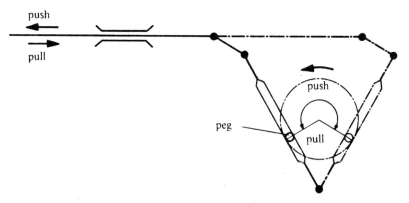

iv) The push movement will take longer than the pull movement (the reason for this is shown on the diagram), and this mechanism is often used when a quick-return motion is required.

To reverse these conditions — i.e. to obtain a slow, large pull movement — reverse the direction of rotation of the peg.

v) The motion of the slider is reciprocating.

Applications
Used in shaping machines.

c) **The cam and follower**

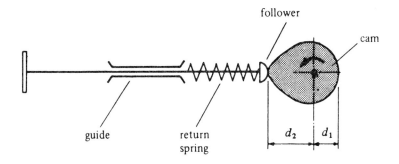

Properties
i) Large push forces can be exerted when the cam pushes the follower. The follower returns either by gravity or by means of a spring. Some of the energy used to rotate the cam is used to compress the spring.

ii) The spring action returns the follower, and therefore the pull forces depend upon the strength of the spring. The pull forces are usually considerably smaller than the push forces, and for this reason the cam follower is normally used to provide push forces only.

iii) Usually used only for small linear movements, otherwise very large cams would be needed.

iv) The shape of the cam determines the characteristics of the linear motion produced.

v) The distance the follower moves is determined by the distance $d_2 - d_1$, as shown in the diagram.

vi) The motion is reciprocating.

Applications

The valve gear in motor cars.

d) The eccentric

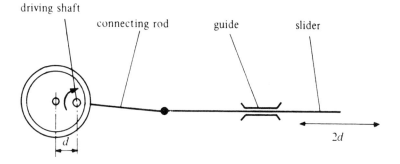

driving shaft

connecting rod   guide   slider

d

$2d$

Properties

i) Similar in certain respects to the cam, but produces powerful push *and* pull forces.

ii) The amount of movement produced is equal to $2d$.

iii) Used for comparatively small linear movement, otherwise the mechanism would be bulky.

iv) The motion is reciprocating.

Applications

The valve gear in steam locomotives.

c) The rack and pinion

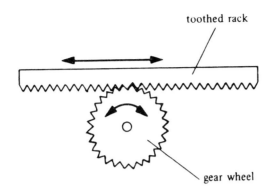

Properties

i) The rack moves continuously in one direction if the gear continues to move in one direction, i.e. the motion is *not* reciprocating but is in one direction only.

ii) Large push *or* pull forces can be exerted.

iii) The motion of the rack is uniform throughout one revolution of the gear, provided that the speed of the gear is constant.

Applications

Some motor-car steering boxes. Hand-operated carriage movement on some lathes.

f) The screw-thread

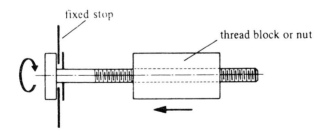

Properties

i) Very large forces can be exerted in both directions.

ii) If the screw rotates continuously in one direction, the threaded block also moves continuously. Movement of the block is reversed by rotating the screw in the opposite direction.

iii) The motion is *not* reciprocating.

Applications

Leadscrew on a lathe. The vice. Some window-opening devices. Some motor-car jacks.

Linear Motion Follow-up 2 ⎯⎯⎯⎯⎯

a) Some factors which could possibly affect the force provided by the piston of a pneumatic cylinder are:
   i)   the position of the piston in the cylinder;
   ii) the pressure of the air supply;
   iii) the piston diameter, which may or may not be the same as the cylinder diameter. (Look carefully at a double-acting cylinder.)

b) Some factors which could possibly affect the force provided by a solenoid are:
   i)   the position and length of armature in the solenoid;
   ii) the current flowing in the solenoid;
   iii) the number of turns on the solenoid;
   iv) the material from which the armature is made. (A coil which carries a current becomes an electromagnet; the magnetic field produced attracts only armatures made from iron, steel, or certain special alloys.)

c) There are basically two kinds of pneumatic cylinder:
   i)   *Single-acting*

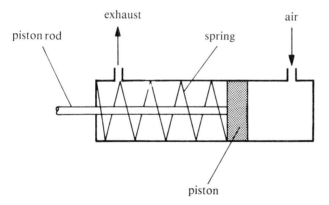

The spring returns the piston when the input air pressure is removed.

ii) *Double-acting*

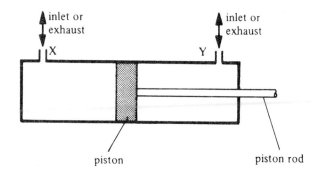

Pressure applied at X forces the piston rod out, and air is expelled from Y.

Pressure applied at Y forces the piston rod in, and air is expelled from X.

Linear Motion Follow-up 3

The statements which follow are based upon actual results obtained from investigations. Your results may not be identical to these, but they should be broadly similar.

3.1 Solenoids

To investigate the relationship between the position of a solenoid armature and the force produced

A graph similar to that shown below should have been obtained, indicating that the force produced varies with the position of the armature.

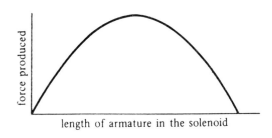

length of armature in the solenoid

In *assignments 3.2* and *3.3*, it is essential that the armature position is constant throughout each experiment. The positioning must be carried out with the current switched on, otherwise the armature will move as soon as the current starts to flow.

3.2 Solenoids

To investigate the relationship between the current flowing in a solenoid and the force produced

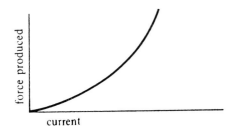

current

The shape of the graph suggests that the force produced is proportional to the square of the current.

This can be verified by plotting the force produced against the square of the current.

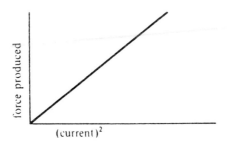

If the result is a straight line, as indicated above, it can be concluded that the force produced is proportional to the square of the current flowing in the solenoid coil.

3.3 Solenoids

To investigate the relationship between the number of turns on the solenoid and the force produced.

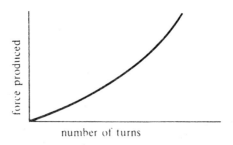

The shape of the graph indicates that a second graph should be drawn to establish a relationship. Note the similarity between this curve and that obtained in 3.2.

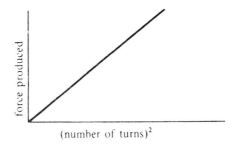

(number of turns)²

If a straight-line graph is obtained, it would appear that the force produced is proportional to the square of the number of turns on the solenoid.

Do your experimental results agree with these?

Summary – solenoids

The force produced on the armature of a solenoid depends upon:
) the position of the armature;
i) the current flowing in the solenoid (experimental results suggest that the force is proportional to the square of the current);
ii) the number of turns on the solenoid coil. (Experimental results suggest that the force is proportional to the square of the number of turns.)

Your teacher will have demonstrated *assignment 3.1* again, but this time with an iron insert screwed into the end of the solenoid.

The properties of the solenoid are modified considerably, as can be seen from the graphs below.

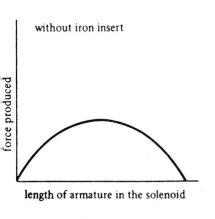

without iron insert

length of armature in the solenoid

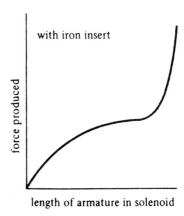

with iron insert

length of armature in solenoid

When the end of the armature nears the iron insert, the intensified magnetic field, due to the presence of the iron, exerts a very large pull force on the armature. The armature is therefore pulled very strongly further into the solenoid and, when in contact with the iron insert, a very large force is required to remove it.

Although a solenoid with an iron insert is capable of producing only moderate pull forces for most of the armature movement, it does provide a very useful 'locking' facility when the armature is fully inserted.

Application

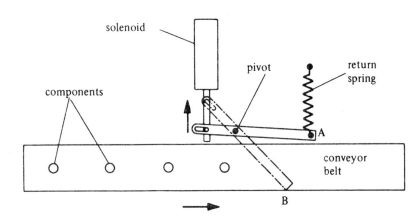

The solenoid is used to operate the lever which removes components from the conveyor belt.

The force produced on the solenoid armature may be just sufficient to overcome the resistance of the return spring. If the components are heavy, the armature may not be able to enter the solenoid once the lever has come into contact with the component.

However, once the lever is in position B, a larger force is required to displace it. Therefore, with such a mechanism, the movement should be completed before the component to be removed comes into contact with the lever.

Note that an electro-magnetic relay is very similar in principle to a solenoid.

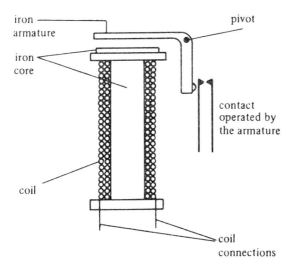

3.4 Pneumatic Cylinders

To investigate the relationship between the position of the piston and the force produced

The position of the piston has no effect upon the force produced by a pneumatic cylinder (provided all other factors are kept constant).

This is easily understood by studying the construction of a pneumatic cylinder.

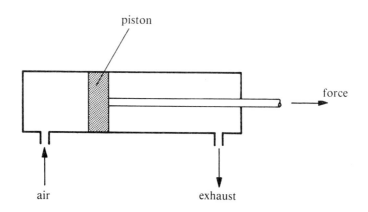

If the pressure of the air supply is 5 bar (which is equivalent to 500 kN/m²) then, for every square metre of piston area the force on the piston is 500 kN. Piston areas are usually smaller than a square metre. If the piston area is 20 mm² the force exerted on the piston, and therefore by the piston rod, is

$$\frac{20}{1\ 000\ 000} \times 500\ 000 = 10 \text{ N}$$

Thus,   force produced = air supply pressure × area of piston

or, more generally   force = pressure × area

The area of the piston cannot change, and, therefore, if the pressure remains constant, the force produced is constant.

A graph of your results, although obviously not necessary, would have the shape shown below.

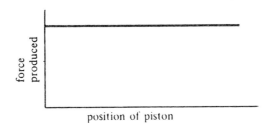

position of piston

As a pneumatic cylinder provides the same force throughout the whole of the piston-rod movement, there is no need to take the piston position into account in future experiments.

3.5 Pneumatic Cylinders

To investigate the relationship between the pressure of the air supply and the force produced

A graph similar to that shown below is usually obtained.

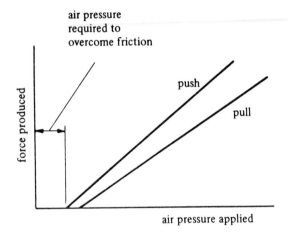

a) A straight-line graph indicates that the force produced is proportional to the pressure applied.

b) The graphs clearly indicate that the push force is always greater than the pull force for any particular pressure.

c) The graphs do not pass through the origin; a certain pressures is required to produce any movement of the piston rod. This pressure is needed to overcome the frictional forces at the air seals A and B.

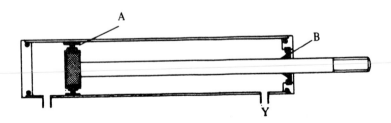

It may be found that the frictional forces are slightly higher when pulling than when pushing. This is probably due to the additional friction produced between the air seal B and the piston rod when the air pressure is applied at Y.

3.6 Pneumatic Cylinders

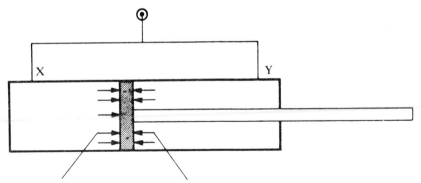

area producing the push force   area producing the pull force

When equal air pressure is applied at both X and Y, the piston rod will move out, but only slowly.

The pressures on each side of the piston are, of course, equal. However, the area of one side of the piston is greater than the other, since the piston rod reduces the effective area.

Therefore the push force will always be greater than the pull force since

force = pressure × area

Example

Cylinder of 18 mm diameter bore (18 mm diameter piston); piston rod 6 mm diameter:

the area producing the push force is the area of the piston

$= \pi \times$ (radius of piston)^2

$= \pi \times 9^2$

$= \pi \times 81$

$= 81\pi$ mm^2

the area producing the pull force is the area of the piston minus the cross-sectional area of the piston rod

$= \pi \times$ (radius of piston)$^2 - \pi$ (radius of piston rod)^2

$= \pi\, 9^2 - \pi \times 3^2$

$= \pi \times 81 - \pi \times 9$

$= 72\pi$ mm^2

Therefore for this cylinder the pull force is approximately 72/81 (8/9) of the push force at any pressure.

3.7 Pneumatic Cylinders _____

To investigate the relationship between the diameter of the piston of a pneumatic cylinder and the force produced

As already established, the force produced by a pneumatic cylinder depends upon the area of the piston and the air pressure applied (neglecting frictional forces).

But the area of the piston is proportional to the square of the radius.

Example

For a 20 mm diameter bore cylinder, the piston area is

$$\pi \times \text{(radius of the piston)}^2 = \pi \times 10^2$$
$$= 100 \, \pi \, mm^2$$

For a 40 mm diameter cylinder, the piston area is

$$\pi \times \text{(radius of the piston)}^2$$
$$= \pi \times 20^2$$
$$= 400 \, \pi \, mm^2$$

Therefore a cylinder twice the diameter of another will provide four times the force at a given pressure; one three times the diameter will produce nine times the force, etc. (neglecting frictional and other losses).

Example

Compare the force produced by a 20 mm and a 40 mm diameter bore cylinder when an air pressure of 5 bar is applied.

Piston area of 20 mm diameter cylinder (when pushing)

$$= \pi \times \text{(radius)}^2$$
$$= \pi \times 10^2 \, mm^2$$
$$= 3.14 \times 100 \, mm^2$$

But force = pressure × area and 5 bar = 500 kN/m²

$$\therefore \qquad \text{force produced} = 500\,000 \times 3.14 \times \frac{100}{1\,000\,000}$$
$$= 50 \times 3.14$$
$$= 157 \, N$$

Piston area of 40 mm diameter cylinder (when pushing)

$$= \pi \times \text{(radius)}^2$$

$$= \pi \times 20^2 \text{ mm}^2$$
$$= 3.14 \times 400 \text{ mm}^2$$

But force = pressure × area

$$\therefore \quad \text{force produced} = 500\ 000 \times 3.14 \times \frac{400}{1\ 000\ 000}$$
$$= 200 \times 3.14$$
$$= 628 \text{ N}$$

i.e. at a given pressure, a pneumatic cylinder of 40 mm diameter produces a force four times greater than does a cylinder of 20 mm diameter.

Summary — pneumatic cylinders

Carefully performed experiments using your pneumatic equipment suggest the following:

i)   The position of the pneumatic cylinder piston does not affect the force produced; a constant force is produced throughout the movement of the piston rod.

ii) The force produced is proportional to the air pressure applied.

iii) The force produced is proportional to the effective cross-sectional area of the piston.

iv) The push force provided by a double-acting cylinder is always greater than the pull force.

v)   A single-acting pneumatic cylinder produces large push forces only.

Applications

As pneumatic cylinders produce large forces for the whole of the movement of the piston rod, they may be used in a wide range of circumstances, e.g. to move, hold, and process a component (see opposite).

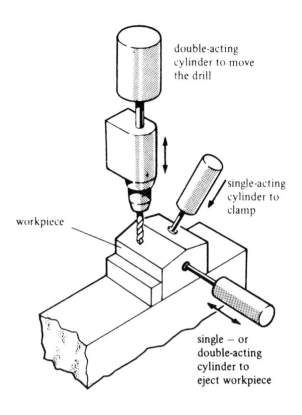

double-acting
cylinder to move
the drill

single-acting
cylinder to
clamp

workpiece

single — or
double-acting
cylinder to
eject workpiece

Pneumatic Control Follow-up 1

Valve operation	Movement of piston rod (in or out)	Speed of movement (fast or slow)	Force produced (small or large)
1 Valve A off, valve B off	No movement	Zero	Piston rod can be moved easily by hand.
2 Valve A on, valve B off	Out	Fast	Large
3 Valve A off, valve B on	In	Fast	Large
4 Valve B on, then valve A on	In, then out	In fast, out slowly	Slow movement provides small force. Piston rod can be pushed back by hand.
5 Valve A on, then valve B on	Out. No charge when valve B is on also.	Out fast	Large force when valve A is on. When valve B is also on, rod can be moved by hand.

The results from 4 and 5, when pressure is applied to both sides of the piston, are explained as follows:

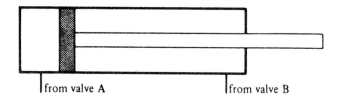

from valve A   from valve B

In 4, when valve B is operated, the piston rod moves in. However, when valve A is also operated, the piston rod moves out slowly, but it can be stopped or pushed back fairly easily by hand.

The pressure on both sides of the piston is the same, but the area providing the OUT movement (valve A) is greater than the area on the IN movement side (valve B) and, because force = pressure × area, the OUT (push) force is slightly higher than the IN (pull) force, and the rod moves slowly outwards.

With valves in position 5, the piston rod moves out due to air pressure from valve A. Air pressure from valve B fails to produce further movement, as the force produced on this side of the piston, due to the smaller area, is less than that produced by the air from valve A.

Only in valve positions 2 and 3 were powerful, fast movements obtained, and, in each of these, one valve was supplying air (ON) and the other was open to exhaust (OFF). Powerful movements are produced only when the air pressure on one side of the piston is much greater than the pressure on the other side. Therefore, to make effective use of the properties of the pneumatic cylinder, one side of the piston must be connected to the pressure supply and the other to exhaust.

By using the valve positions in 2 and 3 alternately, it is possible to make the piston rod oscillate. However, to ensure that the movement of the piston rod is always under control, one side or the other must be supplied with air under pressure. If both valves are open to exhaust, the piston can be moved to any position as indicated in valve position 1.

It is therefore essential, when using a double-acting cylinder, to pressurise one side of the piston and to exhaust the other. If the reverse direction movement is required, these conditions must be reversed.

A 'change-over' valve is obviously required.

You will have observed that when only one of the two lock-down valves is operated, air escapes from the exhaust port of the other valve. Adding the shuttle value to the circuit allows one or other of the two valves to operate the piston movement. Inside a shuttle valve a small ball is blown across the top of the 'tee', blocking the exhaust route. Air can only pass along the vertical part of the tee to the cylinder connection. This route also serves as an exhaust.

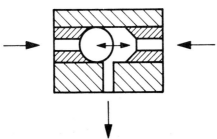

The circuit for continuous oscillation of the piston rod is shown below.

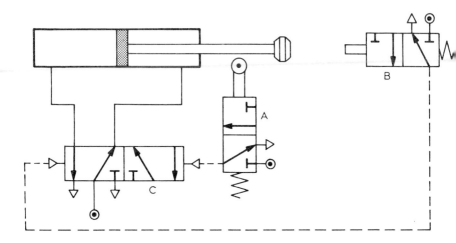

The piston rod oscillates continuously. The double pilot-actuated change-over valve C receives impulses, alternately, from the two piston-rod position-sensing valves A and B.

If these impulses are prevented from reaching the change-over valve by the introduction of two lock-down valves D and E, the piston rod will stop either fully out or retracted, and with the piston still pressurised on one side.

You should have realised that a single-pilot spring-return valve is a monostable device, as it will always return to its stable state when the pilot signal is removed. You will learn about the pneumatic equivalent of a capacitor and how it can be used to increase the time for which the unstable state is maintained. You should also have noted that a double pilot valve is a bistable device and that an astable system was developed when the valve was included in the oscillator circuit.

With valve E ON and valve D OFF, the piston rod will stop OUT.

With valve D ON and valve E OFF, the piston rod will stop RETRACTED.

When valve E is ON and an impulse is applied through valve D, via

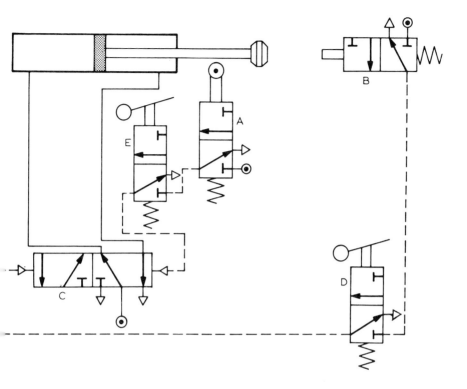

valve B, the piston rod will retract, immediately return, and stop in the same position as it started, thus completing one cycle.

If you were unable to design and test the circuit, connect it up and try it now. More complex circuits using two or more cylinders can be designed using the same control techniques.

Example (See circuit diagram overleaf)

With valve X ON, the circuit will operate continuously as follows:
    piston rod of cylinder A moves OUT
    piston rod of cylinder B moves OUT
    piston rod of cylinder A moves IN
    piston rod of cylinder B moves IN

The sequence is easily altered by interchanging the valve connections. This sequence could be shown as (page 85):

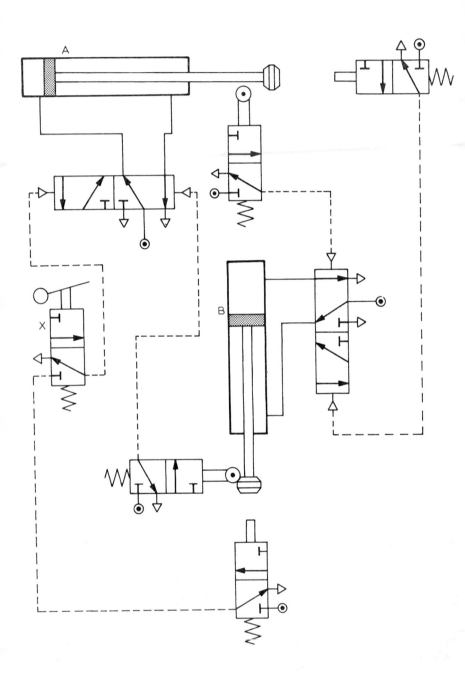

$$\left.\begin{array}{l} A+ \\ B+ \end{array}\right\} \text{ where ' + ' indicates outward movement, and}$$

$$\left.\begin{array}{l} A- \\ B- \end{array}\right\} \text{ where ' – ' indicates inward movement of the piston rod.}$$

With valve X OFF, the continuous cycling will stop with the piston rods in the retracted positions shown.

A short impulse from valve X will start one complete cycle of the operation.

When using a double pilot-operated valve, it is essential that the air pressure is released from one pilot port before pressure is applied at the other, or the valve will not change over.

Example

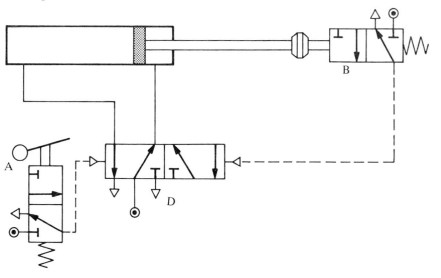

The air pressure from valve B cannot change over the valve D and return the piston rod until valve A is switched OFF.

For this reason, the most reliable way of operating such a valve is by applying impulses to the pilot ports, as practised in the previous oscillating circuits.

The relay bistable unit used in the electrical-switching assignments demands similar conditions. A short-circuit across one pair of green sockets must be removed before a short-circuit at the other pair of sockets has the desired effect.

85

a) **Continuously oscillating double-acting cylinder with speed control using exhaust restrictors**

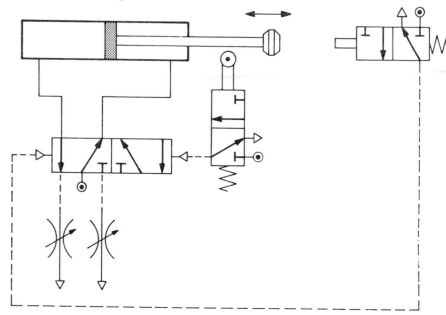

The exhaust restrictor cannot be used to control the speed of the powerful push movement of a single-acting cylinder. It could, however, be used to ensure a slow return of the piston rod, but this has limited application.

Example

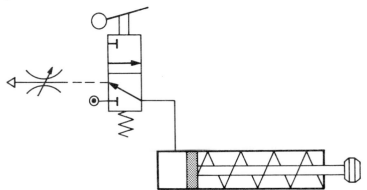

The flow-control valve provides a means of obtaining speed control of a single-acting cylinder. This device limits the volume but not the pressure of the air entering the cylinder, and therefore a slower, but powerful, movement may be obtained. This method is applicable to both single-acting and double-acting cylinders.

b) **Speed control of single-acting cylinder**

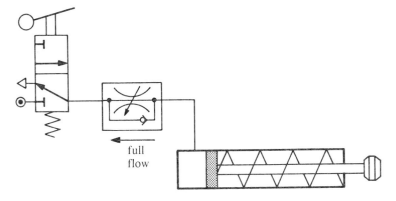

full
flow

c) **Speed control of a double-acting cylinder using one flow-control valve and one exhaust restrictor**

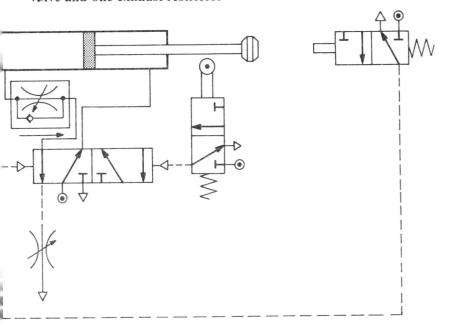

a) An impulse at valve A causes the piston rod to extend, and the cylinder remains pressurised for a short period after the impulse is removed.

The reservoir E is normally connected to the supply, and is charged with air at high pressure. When an impulse is applied at valve A, valve B operates; air from the reservoir E operates the diaphragm-activated valve D and the cylinder is pressurised. The air pressure in the pipe connecting valves B and D falls, due to the presence of exhaust restrictor C, and when a value of approximately 0.2 bar is reached, the diaphragm returns to its unoperated position and the cylinder piston rod retracts.

The size of the reservoir will not influence this circuit to any great extent, as it is in operation only during the application of the impulse. However, adjustment to the exhaust restrictor will affect the length of the delay — the greater the restriction, the longer the delay.

This circuit is not satisfactory for producing a long delay because, if the impulse is very short and the pipe between valve B and valve D is short, only a very small volume of air — that trapped in the pipe — is available to operate valve D, and the pressure will soon fall.

A circuit providing a more effective way of obtaining this result is investigated in *Pneumatic control 5*.

When valve A is switch ON and left ON, the piston rod immediately extends, and after a period of time retracts. (The period of operation may be controlled by the exhaust-restrictor setting and the size of the reservoir.)

The reservoir E is normally connected to the supply, and is charged with air at high pressure. When valve A is switched ON, valve B operates and allows air from the reservoir to operate diaphragm valve D, and the cylinder is pressurised. Since valve B remains operated, the air stored in the reservoir gradually escapes through exhaust restrictor C and, when the pressure falls to approximately 0.2 bar, valve D returns to its unoperated position and the cylinder piston rod retracts.

This circuit therefore converts a continuous input into an output of short duration.

**NOTE: If the pipe connecting valve B to valve D is long, the
limited quantity of air when released from the reservoir will
cause only a moderate rise in pressure in this pipe. For this
reason, the circuit should be used to operate only low-pressure
activated valves, e.g. diaphragm valves.**

The electrical equivalent is:

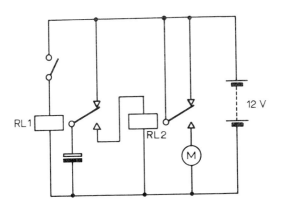

Application of this circuit
The following circuit was discussed in *Pneumatic control follow-up
2;* the piston rod cannot retract until valve X is switched OFF.

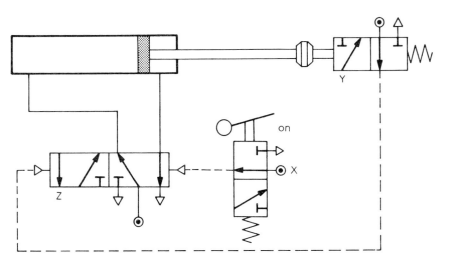

However, if the circuit under investigation is introduced between valve X and valve Z, it is possible for the cylinder to complete one cycle, although valve X remains switched ON.

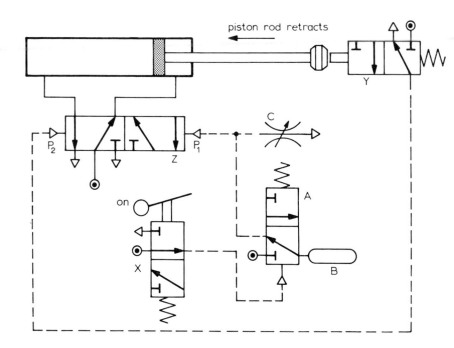

When valve X is switched, the air contained in the reservoir B is released to operate valve Z, and the piston rod extends. The air pressure in the pipe between valves A and Z immediately falls, due to the presence of exhaust restrictor C. When the piston rod operates valve Y, valve Z will change over, because the air supply pressure from valve Y acting at pilot port P2 will be greater than that acting at pilot port P1.

The piston rod will not extend again until valve X is switched OFF (which allows the reservoir to be recharged) and then switched ON a second time.

An application of the circuit is shown opposite:

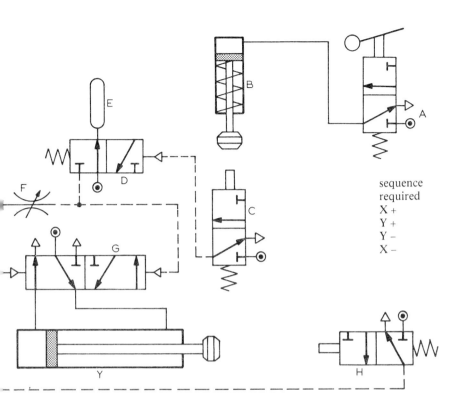

sequence
required
X +
Y +
Y –
X –

When valve A is switched on, the piston rod of cylinder X moves
out and operates valve C, which remains ON. If the output from
valve C is taken directly to bistable G, it will not be possible for
sensor H to change over this valve unless valve C is released.
However, by introducing valve D, reservoir E, and exhaust
restrictor F, the pressure on the pilot of valve G falls as the
reservoir E discharges its air through restrictor F, and valve H can
then change over bistable G to return the piston rod of cylinder Y.
The piston rod of cylinder X will retract when valve A is switched
OFF, thus releasing valve C.

An important safety factor is that valve C must be released before
reservoir E is recharged and valve G can be operated again. This
ensures the correct sequence of operation.

b) Same circuit as (a), but with pipes X and Y interchanged.

Nothing appears to happen until valve A is switched OFF, then the piston rod extends and remains out for a period of time dependent upon the size of the reservoir and the setting of the exhaust restrictor.

When valve A is switched ON, valve B changes over, and the reservoir E is very quickly charged directly from the air supply.

When valve A is switched OFF, valve B returns to its unoperated position, causing the air in the reservoir to operate diaphragm valve D. The presence of exhaust restrictor C causes the air pressure on the diaphragm to drop, which, on falling to approximately 0.2 bar, allows valve D to return to its unoperated position, exhausting the cylinder F.

This circuit may be used to start an operation on the completion of another operation.

Example: the automatic removal of a component when a clamp is released.

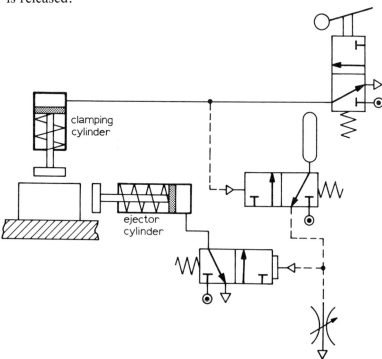

The ejector cannot operate again until the reservoir has been recharged by the operation of the clamping cylinder.

The electrical equivalent of this circuit is:

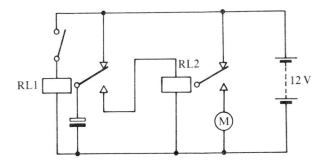

a) When an impulse is applied at valve A, the piston rod extends
and the cylinder remains pressurised for a period of time
dependent upon:

i) the adjustment of the flow-control valve,

ii) the size of the reservoir.

When valve A is switched ON, the reservoir is charged and valve
D operates almost immediately, causing the piston rod of
cylinder E to extend. On switching valve A OFF, the flow-
control valve B restricts the flow of air from reservoir C through
the exhaust of valve A so that valve D remains operated until the
pressure falls to approximately 3 bar.

If valve D is replaced by a diaphragm valve, a much longer delay
will be obtained, as the valve will remain operated until the
pressure has fallen to approximately 0.2 bar.

This circuit can be used to increase the duration of an impulse,
or increase the duration of any operation.

Electrical equivalent

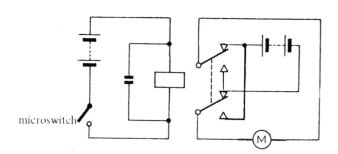

Do you recognise this circuit? You used it in an electrical-
switching assignment to reverse your vehicle away from an
obstruction.

When the microswitch is closed, both the relay and the capacitor
are energised. The relay remains energised for a period after the
switch is opened, held on by the discharging capacitor.

b) An impulse applied at valve A will not operate the circuit. However, if valve A is switched ON and left ON, after a period of time the piston rod of cylinder E suddenly extends.

The DELAY time can be controlled by
i) adjusting the flow-control valve,
ii) changing the size of the reservoir.

When valve A is switched ON, the air pressure between the flow-control valve B and valve D gradually builds up, and, when it reaches a value of about 3 bar, valve D is operated, causing the piston rod of cylinder E to extend.

The cylinder remains pressurised until valve A is switched OFF, when the whole system is immediately exhausted.

This is a very useful circuit because it allows one valve to control two or more cylinders, and ensures that the operations commence in a particular sequence.

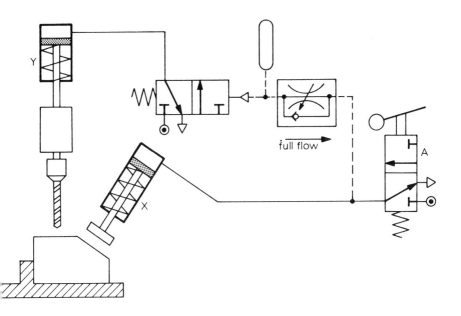

When valve A is switched ON, cylinder X extends immediately, to be followed, after a delay, by cylinder Y.

Electrical equivalent

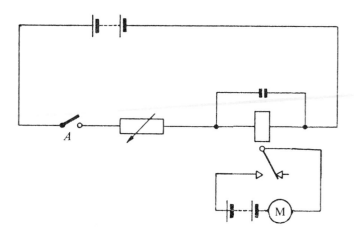

This is not exactly an equivalent circuit because in the pneumatic circuit the piston rod retracts immediately valve A is switched off. In this electrical circuit, the capacitor would hold on the relay for a time after switch A is opened. To produce an identical operation, it would be necessary for the capacitor to rapidly discharge when switch A is opened.

As you have probably realised, the combination of a reservoir and a restrictor in a pneumatic circuit can provide similar results to that of a capacitor and a resistance used together in an electrical circuit.

The two systems are similar in several respects.

To increase the duration of a delay

i) *in the electrical circuit*

Increase the value of the capacitor, and therefore increase the quantity of electrical energy stored.

Increase the resistance of the circuit into which the capacitor discharges its energy, e.g. the relay coil. The larger the resistance, the longer the capacitor takes to discharge its energy.

ii) *in the pneumatic circuit*

Increase the size of the reservoir, and therefore increase the quantity of air stored.

Adjust the restrictor to increase the resistance to the air flow. This slows down the rate of pressure drop within the circuit.

Pneumatic Control Follow-up 6

a) *Sequence of movement*

  X +   the cylinder X clamps the workpiece.

  Y +   the drill moves towards the workpiece slowly, after a period of delay.

  Y –   the drill returns quickly, having produced a hole of the correct depth.

  X –   the clamp releases the workpiece after a period of delay.

b) *Operation*

Pressure from valve A must first reset valve K before valve B can be actuated to operate clamping cylinder X.

Pressure from valve B also reaches valve E through the delay consisting of components C and D.

When valve E operates, reservoir F discharges to operate bistable H, and the pressure of this circuit quickly falls, due to the exhaust restrictor G.

Bistable H causes the drill to move towards the workpiece slowly, and valve J returns the drill when the hole is of the correct depth.

Valve J also operates valve K, which, through another delay (components L and M), operates valve B to release the clamp.

The sequence is then complete.

The next impulse at valve A must reset valve K before valve B can change over and pressurise the clamping cylinder again.

c) *Disadvantage of the circuit*

The operation of the circuit depends to a great extent upon the pressure change, and not the movement of the mechanism. Therefore, if a valve or cylinder failed to function, the operating sequence could be seriously upset, e.g. if the drill did not return on operating valve J, the clamp would still release after a delay, and the drill and workpiece could be damaged.

Because of this disadvantage it would be unwise to use this arrangement.

An alternative solution is now illustrated. This circuit, using mechanically operated valves, would be not only cheaper but also more satisfactory because, if any component failed to function, the sequence of operations would stop immediately.

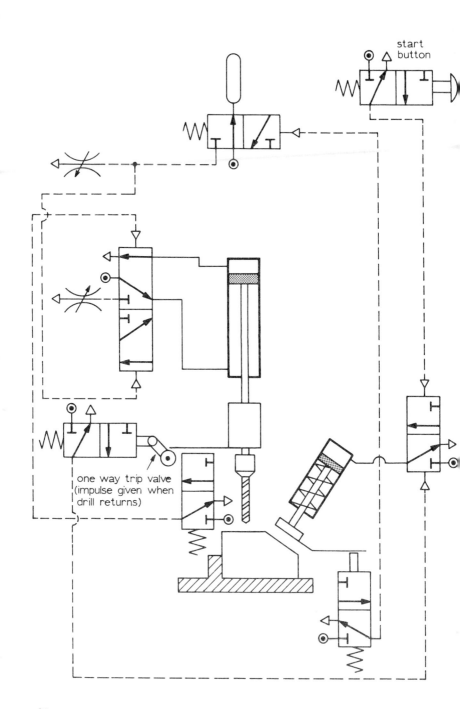

start button

one way trip valve
(impulse given when
drill returns)

98

In this circuit, the completion of one operation causes the next to begin. Therefore the correct sequence is always maintained, although component failure would bring the cycle to a halt.

The main disadvantage of this circuit is that three position-sensing valves are required, and difficulty is often encountered in fitting these in the limited space available on a machine.

Electronics: Resistance Follow-up 1

a) The current taken by your electric motor will depend very much on the exact supply voltage and the type of motor used. When the motor is only slightly 'loaded' (i.e. merely driving the low-friction gear box) the current is unlikely to be greater than 0.5 A (500 mA).

b) With a 25 ohm resistor fitted in series with the motor, the speed of the motor should decrease noticeably. The extra resistance in the circuit will have resulted in a smaller circuit current, and you should also notice that the voltage across the motor terminals is now considerably less. This lower voltage 'across' the motor has the same effect as connecting the motor to a smaller battery.

c) You will also have measured the voltage across the resistor and found that it is quite high. If you add together the two voltages you will notice that the sum is equal to the supply voltage.

When a voltage is applied to a circuit consisting of two series resistances (the motor has resistance), the supply voltage divides itself between the two resistances.

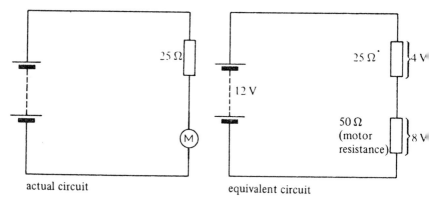

actual circuit   equivalent circuit

The potential differences obtained will depend upon the values of the resistances. In the diagram, the motor is assumed to have a resistance of 50 ohms. This would give a p.d. of 4 volts across the 25 ohm resistor and a p.d. of 8 volts across the 50 ohm motor resistance.

Because the supply voltage divides itself between the resistances, we can call a series resistance arrangement a 'potential-divider' network.

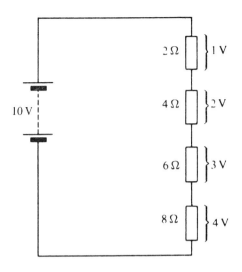

A potential divider can also be obtained by using a variable resistor.

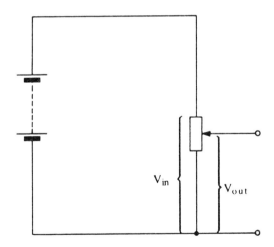

d) You will have calculated the effective resistance of your motor when it is running at maximum speed. Is the p.d. across your motor in agreement with the value you have calculated?

e) You have calculated resistance by dividing the p.d. across a resistance by the current flowing in it.

In symbol form, $R = \dfrac{V}{I}$

where R is the resistance, V is the potential difference, and I is the current.

also $\qquad V = I \times R$

and $\qquad I = \dfrac{V}{R}$

These expressions are derived from a law known as Ohm's law.

f) When *two* 25 ohm resistors are connected in *series*, the effect is the same as when connecting *one* 50 ohm resistor in the circuit, i.e. simply add up the individual resistances in any series circuit to find the total resistance.

Thus $\qquad R = R_1 + R_2 + R_3 + R_4$, etc.

Example

The value of a 1000 ohm, a 500 ohm, a 50 ohm, and a 20 ohm resistor connected in series is

$$R = 1000 + 500 + 50 + 20 \text{ ohm}$$
$$= 1570 \text{ ohm}$$

Electronics: Resistance Follow-up 2

a) The resistance of a tungsten filament lamp varies considerably
with temperature. The resistance is low when the filament is cold
and fairly high when hot. All metals behave like this to a greater
or lesser extent, but the resistance of carbon decreases with
temperature rise.

You have now found two devices whose resistance varies with
the conditions, a filament lamp and an electric motor. Both of
these devices have a very low resistance when first switched on.

The graphs below show the way in which the current changes
with time for a 12 volt 24 watt electric lamp and a 12 volt d.c.
motor with about half of its maximum mechanical load applied.

You can see from these graphs that the 'turn-on' current for
these devices is much larger than the running current.

When designing circuits it is important to remember, that these
current surges can damage devices such as relays and transistors.
The transistor units which you have been using are very robust
and will carry the current surges from any device which you are
likely to use. But you may find problems with project work
where you use other transistors.

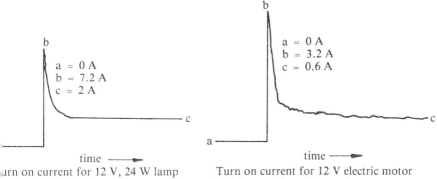

a = 0 A
b = 7.2 A
c = 2 A

a = 0 A
b = 3.2 A
c = 0.6 A

time ⟶

time ⟶

Turn on current for 12 V, 24 W lamp
Total sweep time: 2 seconds

Turn on current for 12 V electric motor
Total sweep time: 2 seconds

Filament lamps often burn out when they are switched on. The
explanation often given for this is that the resistance of the
filament when cold is much lower that when it is hot and that
the relatively large current which flows through the cold
filament causes it to melt. However, on further consideration it
is obvious that before the filament reaches its melting point its

resistance will have risen so that the current flowing through it is back to normal. A better explanation for filament failure is the mechanical shock which is caused by the rapid expansion of the metal filament.

b) The rheostat enables you to vary the speed of the motor continuously rather than in steps, which would be the case if you tried a number of different values of fixed resistor.

You will probably have noticed that, with the rheostat set at maximum resistance, the motor will continue running slowly but will not start from rest. When the motor is first switched on, it consumes a large current compared with its running current, consequently its effective resistance is low when starting compared with the resistance of the rheostat. Most of the voltage is 'dropped across' the rheostat, and only a little (possibly one volt or so) across the motor. One volt is insufficient to enable the motors to start, mainly because of friction.

Try connecting the motor directly to a supply voltage equal to the potential difference you measured across the motor on start. Keep the rheostat at maximum resistance. When the motor is running, its effective resistance is higher, and you will notice that the voltage across the motor terminals is considerably higher than under starting conditions.

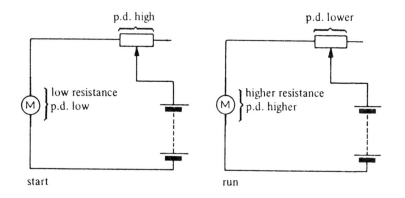

Electronics: Resistance Follow-up 3 _____

a) You will have noticed that the terminal potential difference of a power supply varies with the current consumed. Some power supplies change their terminal potential differences very little over a large range of 'current drain', and we say that these have good 'regulation'. Others vary considerably as the current changes, and we say that these have poor 'regulation'.

In general, as the load current increases, the output voltage of a power supply *decreases*.

b) The terminal voltage of a power supply varies, since all supplies have unavoidable 'internal resistance'. This internal resistance forms a potential divider with the resistance of the device connected to the supply.

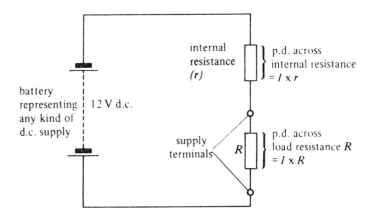

e.g. If the internal resistance is 1 Ω, and the load resistance is 11 Ω,

current in circuit $(I) = \dfrac{V}{R}$

$$= \dfrac{12}{11 + 1} \text{ A}$$

$$= 1 \text{ A}$$

∴ potential difference across internal resistance $= I \times r$

$$= 1 \times 1 \text{ volt}$$

$$= 1 \text{ volt}$$

and potential difference across load $= I \times R$
$$= 1 \times 11 \text{ volts}$$
$$= 11 \text{ volts}$$

This shows that the available battery voltage divides itself between the internal resistance and the load. Since the internal resistance carries the same current as the load, a potential difference appears across it as shown. As a result, the voltage appearing across the terminals of a supply is the difference between the 'off-load' voltage and the p.d. across the internal resistance.

You should now have a clearer idea of the term 'nominal' when you have used a supply of nominal 12 volts d.c. Since the output voltage from any power supply will vary to a greater or lesser extent with the current consumed, the output voltage will be exactly 12 volts at only one particular current.

c) You will have noticed that, under certain conditions, conductors overheat. In this sense a conductor is any device which carries an electric current, e.g. copper wire, resistors, and solenoids. It is obviously desirable to be able to prevent overheating, and the following will help you to determine whether or not a circuit component will overheat.

d) The rate of heat generation within a resistor (or any device carrying a current) is directly proportional to the potential difference across it and the current flowing in it.

i.e.   heat $\propto$ voltage $\times$ current   ($\propto$ means 'is proportional to')

The voltage multiplied by the current is therefore the rate of expenditure of energy within a resistor. This rate of energy expenditure (also called the rate of energy dissipation) is measured in watts (if the potential difference is measured in volts and the current in amperes).

i.e. $V \times I$ = rate of energy dissipation (in watts)

Rate of energy expenditure is also often called power, therefore:

$V \times I$ = power (in watts)

NOTE: 1 watt is equivalent to an energy dissipation of 1 joule per second.

Example

A resistor has a p.d. of 10 volts across it, and carries a current of 1 ampere. What must be the minimum power rating of the resistor?

Power = $V \times I$ watts (VI watts)

= 10×1 watts

= 10 watts

∴ The resistor must have a power rating of at least 10 watts.

e) The 25 ohm and 50 ohm wire-wound resistors you used in your tests were rated at about 5 watts (5 W). We can calculate if the 5 W rating was suitable when a 50 ohm resistor was connected to a 12 volt supply.

Current in circuit (I) = $\dfrac{V}{R}$

= $\dfrac{12}{50}$ A

= 0.24 A

NOTE: 0.24 A can also be written as 240 mA. mA is an abbreviation for milliamperes, 1 milliampere being 1/1000 ampere.

i.e.   1A = 1000 mA

Now p.d. across 50 Ω resistor = 12 V

current (I) = 0.24 A

∴ power = 12×0.24 W

= 2.88 W

The 50 ohm resistors you used are therefore of an adequate power rating. *Now try to calculate the necessary power rating for your 25 ohm wire-wound resistor when you fitted it across the 12 volt supply.* Is its 5 watt rating sufficient?

NOTE: Powers of less than 1 watt are often expressed in milliwatts.

e.g. 0.5 W = 500 mW

f) We can calculate power rating in three ways:
   i) $V \times I$ = power in watts   as you have already learned.
   ii) Since $V = I \times R$, we can substitute $I \times R$ for V:
$$V \times I = \text{power in watts}$$
$$\therefore \quad I \times R \times I = \text{power in watts}$$
$$\therefore \quad I^2 R = \text{power in watts}$$

   iii) Similarly we can substitute for I, since $I = \dfrac{V}{R}$:

$$V \times I = \text{power in watts}$$
$$\therefore \quad V \times \frac{V}{R} = \text{power in watts}$$
$$\therefore \quad \frac{V^2}{R} = \text{power in watts}$$

g) The 10 Ω carbon resistor you used was only of about ½ watt
   rating, and therefore burned out very quickly.
   What wattage should this carbon resistor have been?
   Why did the 10 000 Ω ½ W resistor not burn out? Now work out
   what the lowest power rating would be for a 10 000 Ω resistance
   placed across a 12 volt supply.

   You can use $W = I^2 R$
       or $W = V \times I$
       or $W = \dfrac{V^2}{R}$

   to calculate this value. If you use all three expressions, you will
   find that the same wattage rating is obtained each time.

h) When using photocells, whose resistance changes with the light
   falling upon them, it is important that precautions are taken to
   limit the current in the cell when it is well illuminated, otherwise
   the power dissipation of the cell may be exceeded. The
   photocells you have use have a power rating of about 250 mW.
   If this is exceeded, the device will be damaged beyond repair.
   Discuss this particular problem with your teacher.

a) You should have discovered that, when two identical-value resistors are connected in parallel, the effective resistance is *one half* of the value of one of them. A parallel pair of 50 ohm resistors, therefore, has an effective resistance of 25 ohm.

If at any time you require a resistor of a particular value, and it is not available, a pair of resistors of twice the value will do equally well provided that there is sufficient space to fit them in parallel.

If you require a 100 ohm 10 W resistor, two 200 ohm 5 W resistors in parallel are equivalent. Each 200 ohm resistor need be only 5 W rating, since the current divides equally between the two resistors and therefore the heat generated is shared.

Example

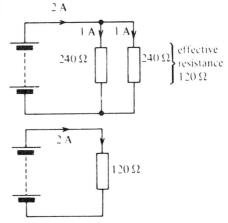

i) Power dissipated in each
    240 Ω resistor
        $= I^2 R$ watts
        $= 1 \times 1 \times 240$
        $= 240$ W

ii) Power dissipated in
    120 Ω resistor
        $= I^2 R$ watts
        $= 2 \times 2 \times 120$ watts
        $= 480$ W

A similar argument applies to series resistors.

Example

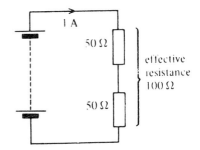

i) Power dissipated in each
    50 Ω resistor
        $= I^2 R$ watts
        $= 1 \times 1 \times 50$ watts
        $= 50$ W

ii) Power dissipated in
100 Ω resistor
$= I^2 R$
$= 1 \times 1 \times 100$ watts
$= 100$ W

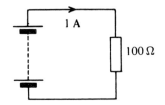

b) When two unequal value resistors are connected in parallel, you will have found that the effective resistance is always *less than the lower value* resistor in the combination.

c) In general, to calculate the effective value of any number of resistors in parallel, we can use the following relationship:

$$\frac{1}{R} = \frac{1}{R_1} + \frac{1}{R_2} + \frac{1}{R_3} + \frac{1}{R_4}, \text{ etc.}$$

Example

Calculate the effective resistance of a 10 ohm and a 20 ohm resistance connected in parallel.

$$\frac{1}{R} = \frac{1}{10} + \frac{1}{20} = \frac{3}{20}$$
$$\therefore R = \frac{20}{3}$$
$$= 6.6 \text{ ohms}$$

Electronics: Rectification Follow-up 1 _____

a) The transformer consists basically of two or more coils of wire
wound one on top of the other. If an alternating voltage is
supplied to the primary coil, an alternating voltage appears
across each of the secondary coils. The size of the voltage at any
secondary depends on the turns ratio between the primary and
the secondary. If the primary has 1000 turns and the secondary
200 turns, then the voltage produced at the secondary will be
200/1000 = 1/5th of that supplied to the primary. Thus, if 250
volts were supplied to the primary, 50 volts would be available
at the secondary. The primary and secondaries are 'coupled'
together by a core of laminated soft iron or other suitable alloy.
Examine a commercial transformer.

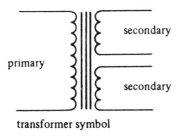

transformer symbol

Note that the energy output from the transformer can never be
greater than the energy input. However, it is a very efficient
device, and for most purposes we can assume 100% efficiency
and we will, therefore, get approximately the same energy out as
we put in.

Power is the rate of expenditure of energy, and therefore the
power input must equal the power output.

i.e. power input ($V \times I$) watts = power output ($V \times I$) watts.

This expression shows that if you step-up a voltage in a
transformer, say from 250 volts to 1000 volts, then the primary
current will be 1000/250 = 4 times the secondary current. In
step-down transformers, the primary current will be less than the
secondary current.

Example

Suppose you 'consume' 2 amperes from your 12 volts
transformer secondary, when the mains supply is 240 volts, then
the primary current will be

$$\frac{12}{240} \times 2 \text{ amperes } = 0.1 \text{ amperes}$$

b) When your motor is connected to a 12 volt a.c. supply, the armature vibrates. Vibration can be considered as an oscillation, which suggests that the motor first tries to go in one direction and then in the other. An alternating current is a current which flows first in one direction round the circuit and then in the other. The mains a.c. changes its direction 50 times per second, which means that the polarity of the supply terminals must be changing every 1/50 th second. Your motor cannot possibly reverse itself 50 times per second, owing to the 'inertia' of its moving parts; it therefore vibrates.

Your oscilloscope trace on a.c. should look something like this:

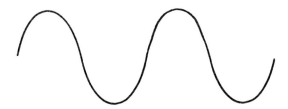

c) When you connected a diode into the motor circuit, you used the following circuit:

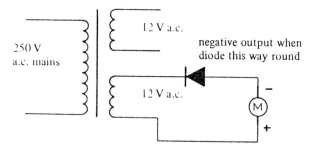

The inclusion of the diode enabled the motor to run. When you reverse the diode, you will have noticed that the motor ran in the opposite direction. These observations show that the current must have been flowing in one direction only, and that to reverse the current flow the diode must be reversed. The diode must, therefore, convert the a.c. to d.c. (direct current — a current which flows in one direction in a circuit).

d) Your oscilloscope traces will probably have looked something like this when you observed the voltage across the motor.

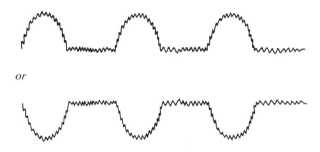

or

With a resistor replacing the motor, the traces would be clearer:

or

e) The diode changes the a.c. to d.c. by removing the top or bottom part of the a.c. waveform.

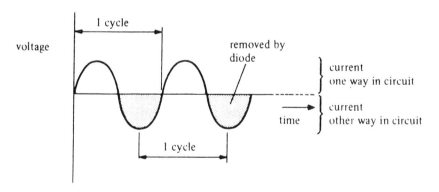

A device which converts a.c. to d.c. is called a 'rectifier'. In this example, one half of the waveform is removed completely, resulting in bursts of d.c. every 1/50 th second. This kind of rectification is called '*half-wave rectification*'.

NOTE: A frequency of one cycle per second is called one hertz (Hz). Normally a.c. mains supply completes 50 cycles every second, i.e. its frequency is 50 Hz.

113

f) The diode conducts in one direction only. If the anode receives a positive voltage or the cathode receives a negative one, it conducts; but if the anode goes negative and the cathode goes positive it does not conduct. You will probably have concluded this from your tests.

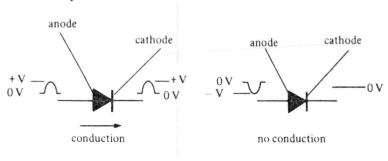

conduction · no conduction

Since the diode receives alternate positive and negative potentials (voltages), it conducts only every half cycle of the input.

The diodes you have used are made from the non-metal silicon containing carefully controlled quantities of impurities. These impurities produce the unidirectional property of the diode. Germanium is also used, but these diodes are more easily damaged by heat than silicon diodes. Ask your teacher to show you some other examples of these *semiconductor* diodes, and notice that they vary considerably in size. All diodes have a maximum current rating; take care not to exceed this rating otherwise your diode will overheat and will be damaged.

g) **Diode testing**

Connect a diode across the terminals of an ohmmeter. Reverse the diode. A good diode should show low resistance in one direction and very high resistance in the other.

Electronics: Rectification Follow-up 2

a) The output from your full-wave rectifier should have looked
something like this:

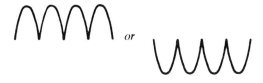

In the half-wave rectifier, the single diode allows the current to
flow during only one half of the cycle. The second diode in the
full-wave rectifier in effect inverts one half cycle of the input,
the result being that both half cycles appear at the output
terminals in a modified form.

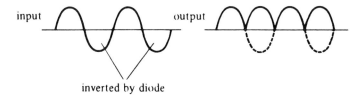

b) When you examined the waveforms at the extreme ends of the
transformer secondary, you will have noticed that they are in
'antiphase', i.e. when one is going positive the other is going
negative, and vice versa.

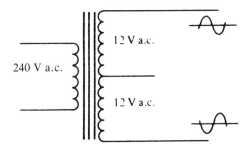

The arrangement allows the diodes to conduct alternately every
half cycle of input.

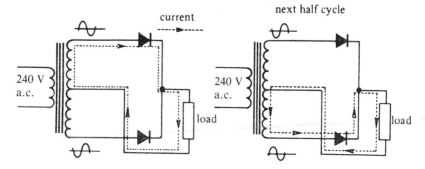

c) The 'bridge' circuit also makes use of both half cycles of input, giving a waveform identical to that which you obtained from the full-wave rectification circuit.

d) The main differences between the circuit components of the full-wave rectifier and the 'bridge' are

 i) a 'bridge' circuit requires only one 12 volt winding, whereas the full-wave circuit requires two (wired as a 24 volt centre-tapped secondary);

 ii) four diodes are required in the bridge circuit and only two in the full-wave rectifier.

e) Since the motor receives pulses every 1/100 th second (100 per second from a full-wave rectifier, it will run at a faster speed than when it runs from a half-wave rectifier receiving pulses every 1/50 th second (50 per second).

f) When a load is placed on an electric motor, the motor has to do more work per second to overcome the loading force, i.e. it must convert more electrical energy into mechanical energy in a given time. This rate of consumption of energy is known as 'power'.

Now power = $V \times I$ watts

Therefore, if more electrical power is to be delivered to the motor, the *current must increase*, since the voltage is constant at 12 volts.

Thus, when you 'loaded' the motor by using your finger as a brake, you should have noticed that the current increased in the circuit.

Electronics: Rectification Follow-up 3

a) When you fitted a capacitor across the rectifier output, your motor will probably have increased in speed. The oscilloscope trace will show that the output from the rectifier circuit to the motor is less 'bumpy' (smoother) when the capacitor is fitted.

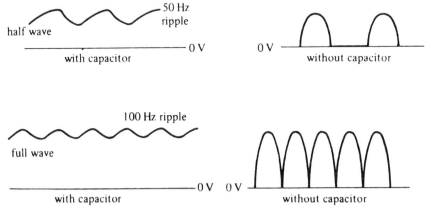

The motor increases in speed because the average output voltage increases when a capacitor is fitted. The output voltage without a capacitor is the average voltage of the pulses, which is lower than the average voltage of the smoothed output — the d.c. plus its ripple.

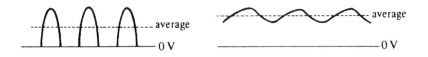

b) If you measure the 'peak' output voltage of a 12 volt a.c. supply on an oscilloscope (remember that an oscilloscope is very useful for measuring voltage), you will find that it is about 16.8 volts. The peak voltage is about 1.4 times the voltage you read on an a.c. voltmeter. When a capacitor is fitted across the output of a rectifier circuit, using a 12 volt a.c. input, the capacitor charges to between 16 and 17 volts and remains charged if no current is drawn from it by an external circuit.

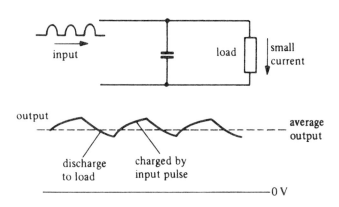

without capacitor   0 V

16.8 V

0 V

with capacitor – no current flow

If a small current is drawn by a load, the capacitor partially discharges before it is again recharged by the next input pulse.

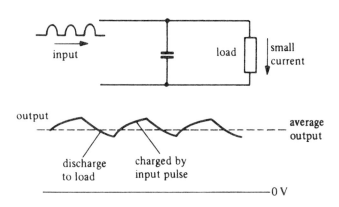

input

load

small current

output

average output

discharge to load

charged by input pulse

0 V

If a larger current is drawn by the load, the capacitor discharges further before it receives its next pulse to charge it up again.

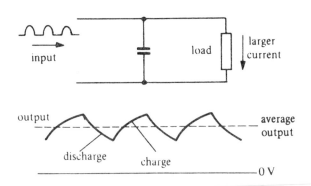

input

load

larger current

output

average output

discharge

charge

0 V

Examination of these waveforms shows that, as the load current increases, the ripple increases and therefore the output voltage falls.

Full-wave rectifiers fitted with a capacitor behave in a similar way, but, since the capacitor is charged at twice the rate of that of half-wave rectification, the output voltage is reduced much less as load current increases, and the ripple is not as pronounced.

For a smooth output from a half-wave rectifier, the capacitor must in general be larger than is necessary for full-waves as explained above. However, the value of capacitor used in any rectification circuit will depend on the currents likely to be consumed and the ripple that can be tolerated. If large currents are required, then a larger capacitor must be used, since it can store a larger charge and will not discharge to too low a voltage before it receives an input pulse. A capacitor used for the above purposes is often called a 'reservoir' capacitor.

Summary

Rectification	regulation	value of capacitor	waveform without capacitor	ripple frequency
Half-wave	poor	higher than full-wave		50 Hz
Full-wave	good	lower than half-wave		100 Hz
Bridge	good	approx same as full-wave		100 Hz

Whilst your investigations will have indicated little difference between a full-wave and a bridge rectifier you should note the following.

a) Bridge systems are generally cheaper to build because silicon diodes cost very little whereas transformers are expensive. The transformer required for a bridge circuit is simpler and cheaper to make than the split-phase units required for full-wave rectification.

b) High-power diodes can produce quite large voltage drops (2 to 3 volts) in circuits. At high currents this can represent a large power loss. In high-current, low-voltage systems a split-phase full-wave rectifier system is preferred to a bridge as the current only flows through one diode (not two). This improves the regulation and efficiency of the system.

Electronics: The Transistor Follow-up 1 ⎓

a) Most modern transistors are made from the element silicon. Germanium, from which all early transistors were made, is rarely used these days except for a few special purposes where its low base-emitter turn-on voltage has advantages.

Silicon can function satisfactorily at very high temperatures (up to 150°C) and can be manufactured to work with voltages between the collector and the emitter as high as several kilovolts.

There are two main types of germanium or silicon transistor: *pnp types* and *npn types*. Your transistor unit contains a silicon npn type. The symbols for the two types are shown below.

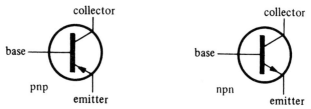

The direction of the arrow in the emitter indicates the type. For everyday usage, both types are interchangeable *but the battery connections must be reversed if an npn type is replaced by a pnp type transistor.*

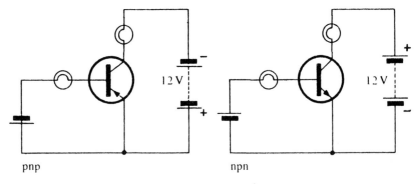

Note that the npn type requires the positive of the supply to be connected to the collector and the negative to the emitter, whereas the pnp type requires the negative of the supply to the collector and the positive to the emitter. **Damage to the transistor may result if you do not use the correct collector-emitter polarity.**

b) The aluminium plate is used as a 'heat sink' for the transistor. The plate provides a larger surface area to allow heat to escape to the atmosphere, thus preventing the temperature of the transistor from rising to too high a value.

c) Your transistor-resistance table should look something like this:

NOTE: Results for a pnp transistor have been included as you may find this information useful in the future.

Test		Resistance (npn transistor)	Resistance (pnp transistor)
Negative terminal of ohmmeter	Positive terminal of ohmmeter		
emitter	base	high	low
base	emitter	low	high
base	collector	low	high
collector	base	high	low
emitter	collector	high	high
collector	emitter	high	high

When you attempted to test the transistor, you will have discovered, as shown in the table, that the resistance measured depends upon which ohmmeter lead is connected to a pair of terminals. Your table should, therefore, have taken this into account in a suitable manner, e.g. as shown above: 'positive terminal of ohmmeter' etc.

Keep this in mind should you meet similar effects elsewhere.

Note that emitter-base and base-collector behave like diodes in some respects (your test with the diode unit will have confirmed this) — but not entirely. A transistor behaves somewhat as though it consists of two diodes 'back to back'.

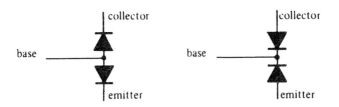

Which diagram represents pnp and which npn?

An important point to note is that usually the terminal of an ohmmeter marked 'positive' is the *negative* terminal of the internal battery; the terminal marked 'negative' is the positive terminal of the internal battery.

Remember that the meter movement is used mainly to measure currents and voltages, and is, therefore, connected in this way.

d) When you set up the following circuit, you will have found that the lamp did not glow (if it did light, you must have been using a faulty transistor). There must therefore be insufficient collector-emitter current to light the lamp.

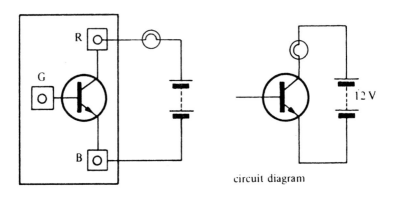

circuit diagram

When the collector is negative, the resistance between the collector and the emitter is high, and therefore limits the circuit current to a value below that required to even partially illuminate the lamp. (Refer to the resistance table.)

e) When you set up the following circuit, lamp A should have been illuminated but not lamp B.

Reversing the 1.5 volt cell extinguishes lamp A. From this one can conclude that the collector-emitter of a npn transistor passes current (has low resistance) only when the base is positive compared with its emitter. If the base is more negative than the emitter, the current between collector and emitter is insufficient to light lamp A.

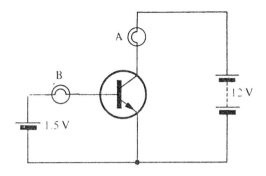

f) Your ammeter readings will have indicated to you that the base current is considerably less than the collector current. The base current was much lower than that needed to illuminate lamp B. The collector-emitter current may have been between 20 and 50 times greater than the base current.

In your experiments, you may have noticed a similarity between the actions of the transistor and the electromagnetic relay, i.e. both act as switches.

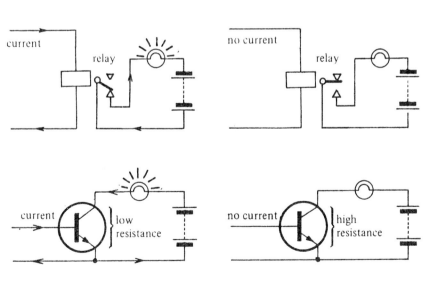

Electronics: The Transistor Follow-up 2 ⸻

a) Your circuit diagram — which includes two power supplies, a lamp, and a transistor — should look something like this:

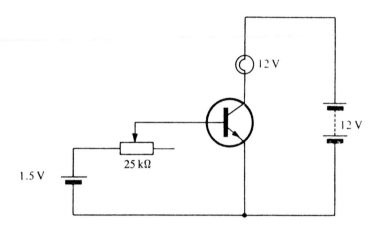

b) When the variable resistor provides high resistance (25 kilohm), the base current is low. If we assume the base-emitter resistance is negligible compared with the 25 kilohm resistance, the base current is:

$$I = \frac{V}{R}$$

$$= \frac{1.5}{25\ 000}\ \text{A} = \frac{1500}{25\ 000}\ \text{mA}$$

$$= \frac{1\ 500\ 000}{25\ 000}\ \mu\text{A} \quad (1000\ \mu\text{A} = 1\ \text{mA})$$

$$= 60\ \mu\text{A}$$

This small base current gives a larger collector-emitter current, but it is insufficient to light the lamp.

When you decreased the resistance in the base circuit, the base current increased, and this produced an increased collector current. At a certain value of collector current the lamp glows, and increases in brightness as the base current increases. Your circuit could be regarded as a 'dimmer' arrangement. The small base current controls a larger collector current to the lamp.

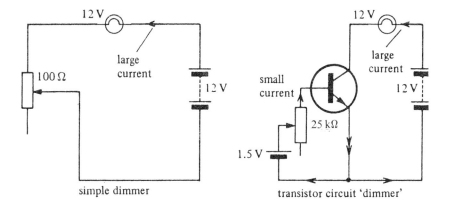

simple dimmer

transistor circuit 'dimmer'

Note that in the *simple dimmer* the large lamp current flows in the variable resistor. The value of the variable resistor must therefore be small to produce 'smooth' control of the lamp illumination.

Make sure that you understand the last statement and, if necessary, make some calculations. Remember that the lamp and rheostat form a potential divider across the supply. (NOTE: A 12 volt 2.2 watt lamp requires a current of $I = W/V = 2.2/12 A = 183$ mA.)

c) i) In *Transistor assignment 1* you learned that a transistor is, in some respects, similar to a relay, since they can both act as switches.

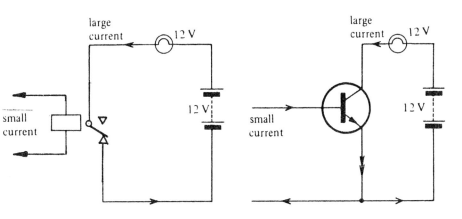

This assignment has also shown you that there is a second similarity. Just as a relay enables large currents to be switched by small currents, the transistor can perform a similar function.

When you replaced the lamp by a motor, you should have been able to control the motor speed by adjusting the variable resistor in the base circuit.

ii) The relationship betwen the base current and the collector current for a transistor is known as its d.c. current again. The symbol h_{FE} is used to indicate this 'current gain'.

$$h_{FE} = \frac{\text{collector-emitter current}}{\text{base-emitter current}}$$

e.g. a base current of 50 μA in a transistor gives a collector current of 1 mA; what is the d.c. current gain of the transistor?

$$h_{FE} = \frac{I_C}{I_B}$$
$$= \frac{1000 \ \mu A}{50 \ \mu A} \quad (1000 \ \mu A = 1 \ mA)$$
$$= 20$$

The current gain of transistors varies considerably, typical values being between 10 and 200.

d) You will have found that speed control of a motor, and lamp control, is possible with the following circuit:

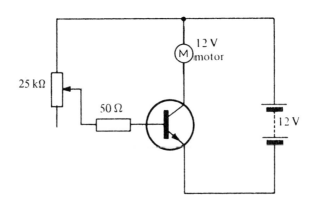

e) You will probably have discovered that the motor speed control is a little better using a transistor compared with simple rheostat control, but at low speeds the motor still provides only a small force at its output shaft. You will probably have been able to stop the motor easily at low speeds.

f) Your circuit diagram, using a photocell and motor, should look something like this:

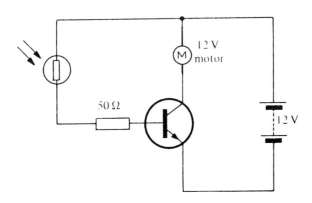

Your explanation of the circuit operation should include these facts:

i)   When no light falls on the photocell, its resistance is high. This gives only a small base current. The small base current provides a larger collector current, but insufficient to start the motor.

ii) When a little light falls on the cell, its resistance falls, the base current increases, and, therefore, so does the collector current. The motor starts but runs slowly.

iii) The fully illuminated cell has a low resistance which gives a large base current. An even larger collector current is produced, and the motor runs at full speed.

a) You should have calculated your transistor current gain as:

$$h_{FE} = \frac{I_C}{I_B} \quad \text{or} \quad \frac{\text{collector current}}{\text{base current}}$$

This method gives only an approximation, but it is accurate enough for most purposes. Did the three sets of readings give the same calculated value for h_{FE}?

Normally, current gain is quoted at a collector current of 1 mA, but at collector currents near 1 mA the value is approximately the same. Note, however, that current gain does change to some extent with collector current.

Your table should look something like this:

Rheostat setting	Base voltage (volts)	State of lamp
Minimum	0	off
10%	0.6	Half bright
25%	0.65	Bright
50%	0.7	Brighter
75%	0.71	Full brightness
Maximum	0.75	Full brightness

You will see that the transistor behaves in a remarkable way. The base emitter junction voltage does not increase much above about 0.6 volts. With a suitable ammeter in the base circuit you will have noticed that virtually no current flows below a voltage of about 0.6 volts. Above this voltage the current rises very rapidly with small increases in voltage.

The inclusion of the ammeter may have made the measured base voltages somewhat higher than they were before the ammeter was fitted. This is due to a small voltage drop that the ammeter itself introduces.

A graph of base–emitter current against base–emitter voltage would look something like that shown on page 129.

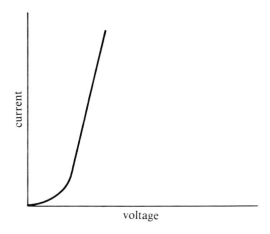

You could have made the investigation of the range between minimum and 10% easier by modifying the circuit.

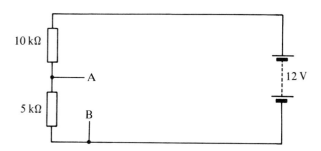

If R is equal to the value of the potentiometer then the same voltage change is produced by twice the rotation of the potentiometer. Larger values of R would increase this effect. To be directly comparable the value of the potentiometer should be changed to keep the total resistance to 26 kilohm.

Your description of a transistor should have included the following.

i) A transistor is a device in which the current flowing in the collector–emitter circuit is controlled by the current flowing in the base–emitter circuit. The collector current is larger, by a factor of between 5 and 200, than the base current.

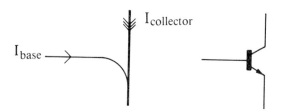

$I_{\text{base}} \longrightarrow$ $I_{\text{collector}}$

ii) In order for any current to flow into the base, the base voltage must exceed about 0.6 volts. (In the case of germanium transistors this is only about 0.2 volts.)

iii) When a transistor passes a current a certain amount of heat is generated. This must be dissipated in some way in order to ensure that the transistor does not overheat.

b) Consider at the circuit below:

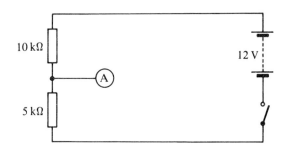

We can calculate the voltage between A and B.

The current through the circuit is given by $I = \dfrac{V}{R}$

$$I = \frac{12}{10\ 000\ +\ 5000}\ \text{A}$$

$$\iota\ = \frac{12}{15\ 000}\ \text{A}$$

If $\dfrac{12}{15\ 000}$ amperes flow through the 5 kilohm resistor the voltage across it $= I \times R$

$$V = \frac{12}{15\ 000} \times 5000$$

$$V = \frac{12}{15} \times 5 = 4\ \text{volts}$$

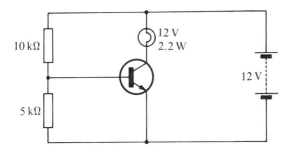

If we now introduce a transistor into the circuit most of the current that would have flowed through the 5 kilohm resistor now flows into the base of the transistor, and of course the voltage across this resistor falls to about 0.6 volts.

We can calculate the current in the base-emitter circuit because we know that all the current in this circuit must flow through the 10 kilohm resistor.

The voltage drop across this 10 kilohm resistor is 12 − 0.6 = 11.4 volts.

The current through this resistor then is:

$$V = \frac{I}{R} = \frac{11.4}{10\ 000}\ \text{A}$$

If the transistor had an h_{FE} of 100 then the collector current would be 100 times this, i.e. about $\frac{1140}{10\ 000} = 0.114$ amperes.

This would make the bulb glow at about sixty per cent of its normal brightness.

If the resistor values were reduced to 1 kilohm and 500 ohms the voltage drop in the base–emitter circuit would again be about 0.6 volts. By the same reasoning as was used above we see that the base current would now be:

$$\frac{11.4}{1000}\ \text{A} = 11.4\ \text{mA}$$

This would give a collector current of about 1.14 A. This current would not in fact flow because the lamp would limit it to about 180 mA. But now the lamp would be fully lit.

If we repeated the calculation again for 100 kilohm and 50 kilohm resistors we would get a collector current of about 11 mA, which would hardly make the lamp glow at all.

In order to design a transistor circuit we must consider a number of points:

i)   The circuit that is to supply current to the base must produce a voltage in excess about 0.6 volts.

ii) The circuit that is to supply current to the base must be able to supply sufficient current.

iii) The required base current can be calculated if we know the expected collector current and the h_{FE} for the transistor.

iv) The transistor must be able to carry the expected collector current. The factors that affect this are complex and the brief data given in most catalogues should be treated with caution. A good rule is to divide the quoted maximum current value by between 5 and 10.

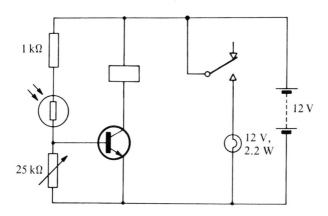

a) You have investigated the above arrangement as a light-operated switch. Your transistor version should be extremely sensitive, and capable of operating reliably over a distance of 1–2 metres.

The rheostat allows you to adjust the sensitivity of the circuit by altering the potential divider.

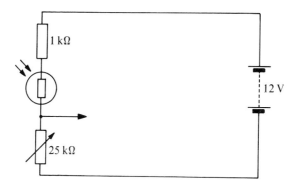

For example, if the rheostat is set to 10 kilohm then the photocell resistance must be less than 190 kilohm in order to raise the potential drop across the variable resistor to 0.6 volts because the photocell, then 1 kilohm resistor and the 10 kilohm resistance of the rheostat form a potential divider. In order to obtain this potential difference, the current flowing through this resistor must be

$$I = \frac{0.6}{10\ 000}\ \text{amperes}$$

The total resistance of the potential divider chain must limit the current from the 12V supply to this value.

From the equation $R = \dfrac{V}{I}$, the total resistance of this potential divider chain must be

$$R = 12 \times \frac{10\ 000}{0.6} = 200\ 000\ \text{ohms}$$

Therefore we will obtain 0.6 volts across the variable resistor, when the resistance of the photocell part of this divider is: 200 − 10 − 1 = 189 kilohms.

If the rheostat is set to 1 kilohm then the photocell resistance must fall below 18 kilohm in order to turn the transistor on.

b) The response time of the transistor system is less than half of the simple system shown below, though the latter is quite fast over short distances.

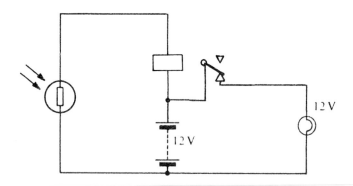

This simple version is not very reliable when the distance between the light source and the photocell is greater than about 0.5 metres.

) A narrow beam of light hitting a photocell enables narrow
objects moving at high speeds to be detected.

A narrow-aperture photocell gives similar results to using a
narrow beam of light.

Your transistor circuit is capable of very fast operation, perhaps
20 switchings per second, but its maximum response rate is
limited by the relay action. It takes time for the relay contacts to
change over, because of their 'inertia'. The transistor you are
using is probably capable of responding to thousands of pulses
per second. The actual number depends upon the response time
of your individual transistor.

) When you reversed the positions of the photocell and the 25
kilohm resistor, you produced the following circuit:

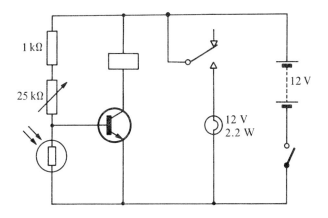

You should have found that this is a more economical circuit
since, when light is falling on the photocell, the transistor is cut
off (relay de-energised), and thus very little current is taken
from the batteries. In the unmodified version, when lights is
hitting the cell — which it does for most of the time — a high
collector current flows, energising the relay.

Logic Follow-up 1

a) Your Truth table for circuit 1 should look like this:

A (input)	B (input)	Z (output)
0	0	0
1	0	0
0	1	0
1	1	1

You get an output only when both inputs A *and* B are present, i.e. only when switches A and B are operated.

The accepted name for this logical arrangement is an 'AND' circuit.

A and B = Z

There is a shorter way to write this equation (Boolean notation):

A . B = Z

b) This is the Truth table for circuit 2:

A (input)	B (input)	Z (output)
0	0	0
0	1	1
1	0	1
1	1	1

There is an output if input A *or* B is present, i.e. if switch A *or* B is operated.

The accepted name for this logical arrangement is an 'OR' circuit.

A or B = Z

which can be written as (Boolean notation):

A + B = Z

(Strictly speaking, it is an 'AND/OR' circuit, since there is an output if A *or* B or A and B are present. For this reason, we call the arrangement an 'INCLUSIVE OR' circuit, since it includes AND and OR — but we usually simplify this to 'OR'.)

c) The Truth table for circuit 3 is:

A (input)	B (input)	Z (output)
0	0	1
1	0	1
0	1	1
1	1	0

Compare this with an AND Truth table. You will probably have noticed a relationship between this Truth table and that of the AND arrangement. The output column of the Truth table for circuit 3 is the binary opposite of that for the AND arrangement, since a 1 is replaced by a 0 and a 0 by a 1.

Output (AND)	Output (circuit 3)
0	1
0	1
0	1
1	0

Because circuit 3 is exactly the opposite (inverse) of an AND arrangement, it is called a 'NOT AND' or simply 'NAND' circuit.

Symbols: $\overline{A \cdot B} = Z$ (the bar indicates inversion)

A 'NAND' arrangement is also sometimes called a 'NEGATED AND.'

The Truth table for circuit 4 is:

A (input)	B (input)	Z (output)
0	0	1
1	0	0
0	1	0
1	1	0

This is exactly the opposite of that for an OR arrangement.

Output (OR)	Output (circuit 4)
0	1
1	0
1	0
1	0

Again, because circuit 4 is exactly the opposite of an OR arrangement it is called a 'NOT OR' or simply a 'NOR' circuit (NEGATED OR).

Symbols: $\overline{A + B} = Z$

a)

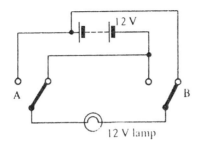

The Truth table for this circuit is:

A (input)	B (input)	Z (output)
0	0	1
1	0	0
0	1	0
1	1	1

The Truth table shows that there is no output unless A and B are *equal*, i.e. they are both 1 or both 0. The arrangement is therefore called a 'LOGICAL EQUIVALENCE' circuit.

Symbols: $A \equiv B = Z$

b)

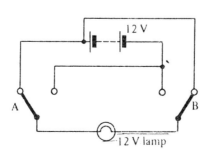

This circuit gives the following Truth table:

A (input)	B (input)	Z (output)
0	0	0
0	1	1
1	0	1
1	1	0

This is purely an OR arrangement, withóut the AND condition. It is therefore called an 'EXCLUSIVE OR' circuit — exclusive of AND. (Do not shorten this to OR, since we do this for the INCLUSIVE OR).

Symbols: $A \neq B = Z$

Logic Follow-up 3

a) Logic devices, whether they are integrated circuits or manual switches connected in different ways are circuit which give an output which is dependent on the state of the inputs.

The logic units which you have investigated in this assignment behave in the same way as the electrical circuits which you looked at in *assignments 1* and *2*. They are all *gates*.

The results you should have obtained are:

i)   AND gate.

A (input)	B (input)	Z (output)
0	0	0
0	1	1
1	0	1
1	1	0

As can be seen from the table *both* inputs, A and B have to be connected to a '1' before the LED glows. For this reason, as explained in *Follow-up 1*, this is known as an AND gate.

ii)   OR gate.

A (input)	B (input)	Z (output)
0	0	0
0	1	1
1	0	1
1	1	1

When *either* A or B is connected to a '1' the LED glows.

For this reason this is known as an OR gate. Note also that the LED glows when A and B are both connected to a '1'. This gate is more accurately called an INCLUSIVE OR gate, since it also *includes* and AND function.

iii) NAND gate

A (input)	B (input)	Z (output)
0	0	1
0	1	1
1	0	1
1	1	0

This gate will light the LED *unless both* A and B are connected to a '1'. This is a NOT AND function the output is exactly the opposite to the AND table results and is called a NAND gate.

iv) NOR gate

A (input)	B (input)	Z (output)
0	0	1
0	1	0
1	0	0
1	1	0

This gate will only allow the LED to glow when *neither* A or B are connected to a '1'. This is called a NOR gate: Then Truth table which we have produced can be further generalised by calling the inputs A, B, C, etc. and the output Z.

In 4 the Truth table obtained is

A (input)	Z (output)
0	1
1	0

The purpose of this is to invert the signal at the input, therefore it is called on INVERTER gate or a NOT gate, because the output is not the same as the input.

b) It is useful to write a logic equation to express a problem. We met these in *Follow-up 1* but they are listed again here.

i) AND   A . B = Z   means A AND B gives Z.

ii) OR   A + B = Z   means A OR B gives Z.

iii) NAND $\overline{A . B}$ = Z   means NOT (A AND B) gives Z i.e. not both A and B

iv) NOR $\overline{A + B}$ = Z   means NOT (A OR B) gives Z i.e. neither A nor B

c) When using logic circuits, it is not correct to assume that if an input is left unconnected, it is a '0'. It is not uncommon for an 'uncommitted' input to assume a logic level '1'. The logic units you have been using do make an uncommited input '1', however other units may produce a '0' or be unstable and keep changing. In practise, whenever you build a logic circuit you should never leave inputs unconnected; either link them to a positive supply voltage (logic '1') or 0 volts (logic '0').

Logic Follow-up 4

In this assignment you will have found that logic gates can be made up with separate components.

a) The first circuit which you investigated should have given you the Truth table of a NOR gate.

If both inputs are connected to '0' then the transistor will be 'turned off' so current can flow via the 1 kΩ resistor to the output.

If however either input is connected to a '1' (i.e. + 12 volts) then current flows into the base of the transistor. The transistor is turned on and effectively connects the output to 0 volts.

The input diodes are required to prevent short circuits if one input is '1' and the other '0'

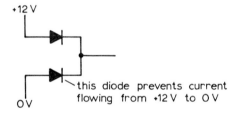

+12 V

this diode prevents current flowing from +12 V to 0 V

0 V

b) By the addition of the extra transistor it becomes an OR gate.

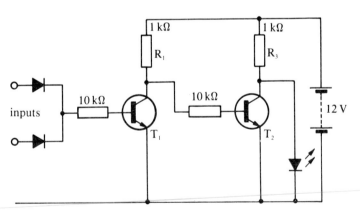

If T_1 is 'turned on' then the current flowing through R_1 is short circuited via T_1 to 0 volts. Thus T_2 must be 'turned off' so current flowing through R_3 can drive the LED at the output.

If T_1 is 'turned off' the current flowing through R_1 can now flow into the base of T_2 so that T_2 'turns on'. Most of the current that was flowing through R_3 into the LED now flows through T_2 and the output is turned off.

c) The third circuit which you investigated is a NAND gate.

In this case if both inputs are connected to a '1' (i.e. $+12$ volts) the two input diodes prevent any current flowing into the gate. But current can flow via the 10 kΩ resistor into the base of the transistor which turns the transistor 'on' and the output 'off'.

If either input is connected to '0' then the current that flowed into the base of the transistor now flows via the input diode to 0 volts, so the transistor turns 'off' and output turns 'on'.

The single diode on the base of the transistor serves an important function. Remember that in order to pass current into the base of a transistor a base-emitter voltage of 0.6 volts is required.

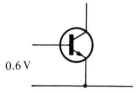

A diode requires the same voltage in order to conduct.

If the input circuit to our gate was

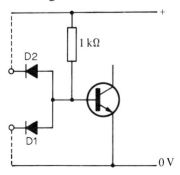

and D_1 was connected to 0 volts (D_2 is shown connected to $+12$ volts but it would not make any difference if this were connected to 0 volts instead) then, because both the diode and transistor base require a voltage of 0.6 V before they conduct, some of the current flowing through the 10 kΩ resistor will still flow into the base of the transistor and it will not turn off. If a diode is connected in series with the base a voltage of 1.2 V is required for current to flow into the transistor so it now flows via the input diode to 0 volts and ensures the transistor turns off.

d) The NAND gate is converted into an AND gate by the second transistor shown in circuit 4.

e) As can be seen in 2 and 4 the addition of the extra transistor inverts the output. (ie it acts as a NOT gate)

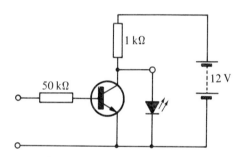

f) The symbols used in this course are quite adequate. However, some books use different symbols. If you have difficulties in this respect, your teacher will help you.

Logic Follow-up 5

It is useful at this stage to remind ourselves of the Truth tables which are obtained from some of the logic gates which we have seen. This will help us to understand the results we get when we combine logic gates together.

NOT

A (input)	Z (output)
0	1
1	0

NOR

A (input)	B (input)	Z (output)
0	0	1
0	1	0
1	0	0
1	1	0

NAND

A (input)	B (input)	Z (output)
0	0	1
0	1	1
1	0	1
1	1	0

a) The Truth table for the arrangement in 1 is

A (input)	B (input)	Z (output)
0	0	0
0	1	1
1	0	1
1	1	1

This truth table is that of an OR gate.

The logic equation for an OR gate is
$$A + B = Z$$

b) The Truth table for this arrangement in 2 is

A (input)	B (input)	Z (output)
0	0	0
0	1	0
1	0	0
1	1	1

This truth table is that of an AND gate.

The logic equation for an AND gate is
$$A . B = Z$$

As can be seen from this it is possible to make a variety of gates with simple combinations of other gates. If you cannot find a NOR gate, for example you could make it from an OR gate and an inverter.

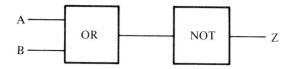

c) The logic equation for the circuit in 3 is
$$(A . \overline{B}) + (\overline{A} . B) = Z$$

Brackets are used so that each section of the equation is readily identifiable.

It is an exclusive OR equation because it eliminates the AND function of the inclusive OR.

You should realise that the circuits you have built only illustrate one possible solution to a problem, the same end result can be achieved in several different ways.

d) The logic equation for the problem set out in 4 is
$$A . B . C . \overline{D} = Z$$

The Truth table is

A	B	C	D	Z
0	0	0	0	0
0	0	0	1	0
0	0	1	0	0
0	0	1	1	0
0	1	0	0	0
0	1	0	1	0
0	1	1	0	0
0	1	1	1	0
1	0	0	0	0
1	0	0	1	0
1	0	1	0	0
1	0	1	1	0
1	1	0	0	0
1	1	0	1	0
1	1	1	0	1
1	1	1	1	0

The block diagram of the logic arrangement is

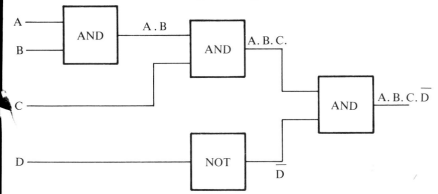

An electrical circuit arrangement using relays could be

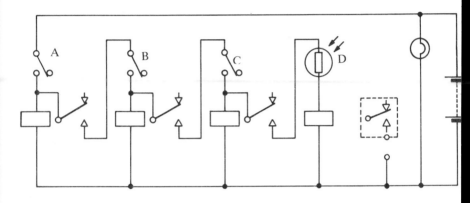